CMOS/TTL Digital
Systems Design

Other McGraw-Hill Reference Books of Interest

Handbooks

Benson • AUDIO ENGINEERING HANDBOOK
Benson • TELEVISION ENGINEERING HANDBOOK
Coombs • PRINTED CIRCUITS HANDBOOK
Coombs • BASIC ELECTRONIC INSTRUMENT HANDBOOK
Croft and Summers • AMERICAN ELECTRICIANS' HANDBOOK
DiGiacomo • VLSI HANDBOOK
Fink and Beaty • STANDARD HANDBOOK FOR ELECTRICAL ENGINEERS
Fink and Christiansen • ELECTRONIC ENGINEERS' HANDBOOK
Harper • HANDBOOK OF ELECTRONIC SYSTEMS DESIGN
Harper • HANDBOOK OF THICK FILM HYBRID MICROELECTRONICS
Harper • HANDBOOK OF WIRING, CABLING, AND INTERCONNECTING FOR ELECTRONICS
Hicks • STANDARD HANDBOOK OF ENGINEERING CALCULATIONS
Inglis • ELECTRONIC COMMUNICATIONS HANDBOOK
Juran and Gryna • QUALITY CONTROL HANDBOOK
Kaufman and Seidman • HANDBOOK OF ELECTRONICS CALCULATIONS
Kurtz • HANDBOOK OF ENGINEERING ECONOMICS
Stout • MICROPROCESSOR APPLICATIONS HANDBOOK
Stout and Kaufman • HANDBOOK OF MICROCIRCUIT DESIGN AND APPLICATION
Stout and Kaufman • HANDBOOK OF OPERATIONAL AMPLIFIER CIRCUIT DESIGN
Tuma • ENGINEERING MATHEMATICS HANDBOOK
Williams • DESIGNER'S HANDBOOK OF INTEGRATED CIRCUITS
Williams and Taylor • ELECTRONIC FILTER DESIGN HANDBOOK

Other

Antognetti and Massobrio • SEMICONDUCTOR DEVICE MODELING WITH SPICE
Antognetti • POWER INTEGRATED CIRCUITS
Elliott • INTEGRATED CIRCUITS FABRICATION TECHNOLOGY
Hecht • THE LASER GUIDEBOOK
Mun • GaAs INTEGRATED CIRCUITS
Siliconix • DESIGNING WITH FIELD-EFFECT TRANSISTORS
Sze • VLSI TECHNOLOGY
Tsui • LSI/VLSI TESTABILITY DESIGN

*For more information about other McGraw-Hill materials,
call 1-800-2-MCGRAW in the United States. In other
countries, call your nearest McGraw-Hill office.*

CMOS/TTL Digital Systems Design

James E. Buchanan
Advisory Engineer
Westinghouse Electric Corporation

Drawings by
Bert D. Buchanan

McGraw-Hill Publishing Company
New York St. Louis San Francisco Auckland Bogotá
Caracas Hamburg London Madrid Mexico
Milan Montreal New Delhi Panama Oklahoma City
Paris San Juan São Paulo Singapore
Sydney Tokyo Toronto

Library of Congress Cataloging-in-Publication Data

Buchanan, James E.
  CMOS/TTL digital systems design / James E. Buchanan; drawings by
Bert D. Buchanan.
     p.   cm.
  ISBN 0-07-008711-3
  1. Metal oxide semiconductors—Design and construction.
2. Transistor-transistor logic circuits—Design and construction.
I. Buchanan, Bert D.   II. Title.
TK7871.99.M4B83   1990
621.381'52—dc20   89-13236
                                                        CIP

1234567890   DOC/DOC   89432109

ISBN 0-07-008711-3

*The editors for this book were Daniel Gonneau and Rita T. Margolies,
the designer was Naomi Auerbach, and the production supervisor was
Suzanne W. Babeuf. It was set in Century Schoolbook. It was composed
by the McGraw-Hill Publishing Company Professional and Reference
Division composition unit.*

Printed and bound by R. R. Donnelley & Sons Company.

*For more information about other McGraw-Hill materials,
call 1-800-2-MCGRAW in the United States. In other
countries, call your nearest McGraw-Hill office.*

Contents

Preface

This book is about the electrical design of high-speed transistor-transistor logic (TTL) and complementary metal oxide semiconductor (CMOS) systems. It is not about logic design. You will not find any discussion of logic simplifications or design of state machines in this book. Instead, it is about the electrical and mechanical environment that logic systems must function in and how to optimize that environment for high-speed TTL and CMOS logic devices. The goal is to help practicing digital designers understand the behavior of high-speed TTL and CMOS devices in a system environment. It is hoped that it will serve as a guide for digital designers on many of the practical aspects of high-speed logic device design and that it will help designers consistently design and build reliable, cost-effective digital hardware.

The material in this book took shape as class notes for an in-house course on digital design guidelines; the course was started in response to the many problems designers were encountering when applying high-speed TTL and CMOS logic devices. In the past, with the older, slower logic families, designers were often able to neglect the electrical system without disastrous results. That is no longer possible. The logic design cannot be separated from the electrical design when advanced TTL or CMOS logic devices are used. The very fast edge rates, very low noise margin, and other nonideal characteristics of high-speed TTL and CMOS devices invite disaster. However, the designer equipped with an understanding of the characteristics of advanced Schottky TTL and advanced CMOS devices and how they react in an interconnection system will be able to design reliable high-speed systems. It is hoped that this book will aid in that understanding.

Throughout this book practical engineering approximations are used rather than rigorous mathematical analysis. The digital design engineer, who must design and build cost-effective equipment in a timely manner, can ill afford to become involved in the solution of elegant equations when the values of the parameters that go into the equations are never known with any degree of certainty. It makes little sense, and serves no useful purpose, to develop elaborate equations and carry them out to many decimal places when the component parameters of the typical digital integrated circuit vary 50 percent or more with process variations and another 50 percent or more with environmental conditions such as temperature and supply-voltage levels. There also may be significant variations in the electrical characteristics of the printed circuit boards or other signal conduction media, and typically none of the parameter variations is well specified. Thus, there is little point to great precision when dealing with digital integrated-circuit device parameters. However, none of the above is intended to imply that digital designers should forego all circuit analysis. Quite the contrary; it is essential that all areas of possible concern be carefully analyzed. Just because a breadboard or a few systems appear to operate correctly, one should never take this as proof of proper design. The great number of possible variations in device and power and signal interconnection system parameters require that a worst-case analysis of all of the relevant system parameters always be undertaken. Failure to analyze properly or even consider some of the most elementary timing or electrical facets of digital systems seems to be an all-too-common trait of many digital (logic) designers. The point is that the analysis must be done in a reasonable and timely manner. The objective in most calculations involving digital devices is to develop a practical engineering estimate of the magnitude of any possible problem. If a simplified approximation shows that a particular factor has little effect on performance, then one can expect that factor to have little impact on circuit operation. If on the other hand, a simplified look at a particular area shows that a potential problem exists, then it is best to take very definitive steps to ensure that the potential source of trouble is eliminated. Some may argue that such an approach is not very exact and may result in "overkill" and is wasteful and expensive. However, practical experience has shown such an approach to be more cost-effective in the long run. Finally, it must be noted that a system with a little excess margin may provide some allowance for those problems that may have been overlooked and not factored into the design, and there are always some of those no matter how good the original analysis.

Book Arrangement

The book is arranged in two parts: material on device characteristics is in Part 1 and design techniques for dealing with the special concerns of high-speed logic devices are presented in Part 2. The designer who has immediate design tasks or some familiarity with advanced Schottky or advanced CMOS devices may prefer to go immediately to specific topics instead of reading the entire book.

Summary of Chapter Content

Chapter 1 stresses that the logic design cannot be separated from the electrical and mechanical design when high-speed logic devices are used. The need for power and signal interconnection systems with wide bandwidth is shown.

Chapter 2 presents a brief overview of the relative performance and specifications of the various TTL and CMOS logic families with the emphasis on the advanced logic families. A number of tables are provided for quick reference and for comparison of characteristics.

Chapter 3 introduces circuit modeling techniques that simplify TTL and CMOS logic circuit analysis. Several circuits are analyzed step by step to help the reader gain an understanding of TTL and CMOS circuit operation.

Chapter 4 defines noise margin, describes the operation of the circuitry that determines noise margin, and discusses the system implications of the noise margins of the various logic families. Tables are provided that list and cross reference the noise margins of the TTL and CMOS logic families that are currently of most interest.

Chapter 5 describes the causes of internal-device and load-dependent transient currents when high-speed logic devices switch. Several examples show the possible magnitude of the switching currents.

Chapter 6 discusses inductance, how to minimize it, and several examples of its detrimental effects in high-speed interconnection.

Chapter 7 describes the need for low-impedance power and ground distribution systems and describes techniques for achieving them. Guidelines for the amount and placement of decoupling capacitors and for calculating loss in power distribution systems are also provided.

Chapter 8 describes the noise rejection capability of synchronous logic designs and the necessity for using synchronous design practices when high-speed TTL or CMOS logic devices are used.

Chapter 9 describes the need for high-quality clock signals and techniques for distributing clock signals. It is stressed that a sound clock distribution system is one of the keys to a reliable high-speed digital system.

Chapter 10 deals with the practical aspects of high-speed digital interconnections.

Chapter 11 includes a review of transmission-line effects and deals with line impedance, signal loading, and crosstalk in high-speed systems.

Chapter 12 provides guidance for determining worst-case device and interconnection delays. A typical board-to-board signal path is analyzed for worst-case signal delay.

Chapter 13 describes techniques for generating and distributing reset or initialization signals. Many critical issues that are often over-looked are discussed.

Chapter 14 covers a neglected topic: the proper connection of unused inputs. The danger of allowing CMOS device inputs to float is noted, and the destructive mechanism is described.

James E. Buchanan

Acknowledgments

Many thanks are due to my fellow engineers at Westinghouse Electric Corporation, Baltimore, Maryland, who contributed to this book through their efforts as leaders or as participants of the in-house digital design guidelines course on which the book is based. Principal contributors include Gray Miller, Michael Juengst, Arden Helland, and Bill Beydler. Gray Miller organized and taught many of the early guideline sessions. Michael Juengst and Arden Helland helped a great deal with Chapters 8, 9, and 11. Jim Hudson and Karl Avellar made the book possible by their advocacy of the guidelines course and by providing the author with the opportunity to acquire the experience on which much of the book is based. The course participants provided invaluable help and encouragement. Their many suggestions to formalize the course material encouraged the author to undertake the book. Special thanks go to Woody Adams, John Baczewski, Al Fitzgerald, and Gregg Mongold for their suggestions; they helped refine the course notes. Also, thanks go to Jim Ross and Carl Nelson for their help with the computer and word processor.

In addition, I want to thank my son Bert for his excellent rendering of the drawings, my wife Beverly and daughter Carol for help with the typing, and Laurie Welms for proofreading the various drafts. All my family are to be thanked for their patience and tolerance during the many evenings and weekends that I spent on the book.

James E. Buchanan

About the Author

James E. Buchanan is an Advisory Engineer at
Westinghouse Electric Corporation's Electronic Systems
Group in Baltimore, Maryland, where he serves as a
Technical Advisor in digital and analog circuit design,
high-speed logic applications, and memory systems. Mr.
Buchanan holds thirteen patents, most of which cover
analog-to-digital conversion and digital-to-analog
conversion techniques. He is also a frequent contributor to
the professional literature in electrical design. Mr.
Buchanan earned his MSEE at the University of
Maryland, and he is a Registered Professional Engineer in
the State of Maryland.

TTL and CMOS
Logic Devices

Introduction

A great deal of care is required for the successful application of advanced Schottky transistor-transistor logic (TTL) and advanced complementary metal oxide semiconductor (CMOS) logic components. Their fast rise times cause crosstalk, transmission-line ringing, power-supply transients, and ground upsets. Unless the electrical interconnection system is designed to minimize the effects of fast edges, there is little likelihood that a sound design will be achieved.

In the past, with the slower logic families, the digital designer was often able to neglect the electrical system without disastrous results. However, that is no longer the case with the new high-speed devices. Yet digital (logic) designers are very often lulled into treating digital components as simple functional logic blocks and neglecting all of the electrical system concerns. It is certainly much easier to deal with simple functional logic blocks, such as the NAND gate shown in Figure 1-1, rather than deal with the actual complex electric circuit, such as shown in Figure 1-2, that exists whenever a high-speed device is applied in a real system. The actual composite circuit, no matter how simple or complex the functional logic unit, also consists of the imperfect interconnection system, the imperfect power and ground system, and the nonideal internal logic device circuitry.

The digital designer must always be aware that the actual electrical environment that surrounds a digital device is similar to that shown surrounding the NAND gate in Figure 1-2. The tendency to view digital circuitry in block form, rather than as a part of a complex electrical network, is one of the leading causes of the design of unreliable digital systems. The basic principles of Ohm's law and electromagnetism cannot be ignored just because the circuitry being applied is digital in nature.

Experience has shown that generally logic errors can be fixed, but a

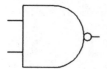

Figure 1-1   Functional block symbol for
a NAND gate.

poorly designed or conceived electrical system may be unfixable. Although it may be unpleasant to change a number of connections to correct a logic error, it can be done. However, if the electrical system is
inadequate, no amount of patching will provide a solid system. One
may be able to hand tune such systems so that they work in the lab
and under some conditions, but they may not work under all the required conditions or they may not work for long. High-speed digital
systems that have an inadequate electrical system or inadequate operating margins may function correctly most of the time, but invari-
ably intermittent or incorrect operation occurs, and often the cause of
the problem is never understood. Very frequently, the problem is
blamed on bad or marginal parts rather than on the real cause of the
problem, an electrical system that is poor or system timing that is not
compatible with worst-case device parameters. In such systems, a
given problem will often go away with a change to another part or another vendor's part. However, in the great majority of such cases, the
part is not the real problem; the problem is that the part is being applied in such a manner that there is very little or no operating margin. Substituting another vendor's part that has a little more margin,

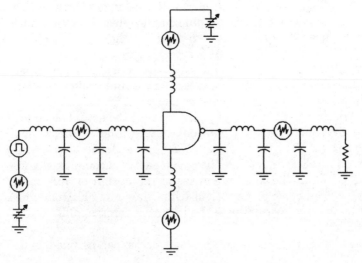

Figure 1-2   NAND gate and surrounding circuitry.

or that is a little faster or slower, corrects the immediate problem but does not fix the basic problem, which constantly reappears in new and varying forms.

High-performance applications amplify the need for care in all aspects of design. Digital designers are forced to push their designs to the limits of the available technology by the constant need to increase processing speed and throughput. The nearer to the limit a system operates, the greater the chance for serious design errors. Because of the pressure for high operating speeds, designers must guard against the use of optimistic typical device parameters to forecast possible system operating speed. Actual in-circuit worst-case device parameters and interconnection delays must be used to determine system operating speed. Designers must not be misled by typical operating specifications for individual components. Under ideal conditions, individual devices may operate at very high speeds. However, when many high-speed devices are interconnected, the interconnections limit the operating speed. With conventional printed circuit board interconnection systems, the practical upper limit for TTL or CMOS systems of any significant size is in the middle to upper 20-MHz clock frequency range, which is much below typical advanced Schottky or advanced CMOS component operating limits.

The fast edge rates of the new logic families are responsible for most of the problems encountered in their application. Signal edge rates, not clock rates, determine the required response of signal and power-distribution systems. The packaging and interconnect techniques that served well for the older, slower, and more forgiving logic families are not adequate for the newer devices. Devices with edge speeds of 1 or 2 ns require high-frequency interconnection systems. For example, when advanced Schottky or advanced CMOS logic devices with switching edges of 1 ns are used, the 3-dB frequency component f of a signal transition t_r is[1,2]

$$f = \frac{0.35}{t_r}$$

$$= \frac{0.35}{1 \times 10^{-9\,s}} = 350 \times 10^6 \text{ Hz}$$

(1-1)

A linear voltage transition, such as a switching edge, also contains a significant amount of higher frequency components. The third harmonic makes up approximately 10 percent of the amplitude of a linear ramp. Thus, an interconnection system that does not have the bandwidth to pass the third harmonic of the 3-dB frequency content of an edge will attenuate the signal to some degree. For the above example, the third harmonic content is approximately 1 GHz (3×350 MHz).

Not all advanced Schottky or advanced CMOS devices have edge rates as fast as 1 ns, but most have edge rates faster than 3.5 ns. A 3.5-ns edge corresponds to 100 MHz [see Table 1-1, which was generated using Equation (1-1)]. Thus, as a minimum, advanced Schottky or advanced CMOS systems need interconnection networks with bandwidths in excess of 100 MHz to prevent serious degradation of signals.

In all cases, to utilize the potential operating speeds of advanced Schottky TTL or advanced CMOS devices, the signal transmission media must be given a great deal of attention. Digital designers must appreciate that they must deal with frequencies that only a short time ago were considered strictly in the domain of the RF designer. When dealing with devices that have edge speeds of 1 or 2 ns and clock rates in the 20-MHz speed range, the functional design and the interconnection and power distribution system can no longer be treated as separate entities. They must be viewed as an integral portion of the overall functional system design. The time delay through the interconnecting wiring cannot be ignored since it may be a significant part of the total system delay. Likewise, waveform distortion due to reflections, resonance phenomena, and signal attenuation due to imperfect transmission media cannot be ignored. To further complicate matters, the loading, load placement, conductor topology, and the conducting and insulating media all influence each of the above. Also, crosstalk between adjacent signals cannot be ignored since it becomes more severe with higher edge speeds, and higher clock frequencies mean that there is less time for it to dampen out. Higher edge speeds and the accompanying large transient currents place much greater demands on the power and ground reference system.

The transient-current generation and noise-tolerance (margin)

TABLE 1-1   Rise Time versus Bandwidth

t_r, rise time, ns	f,* bandwidth, MHz
1.00	350
1.50	233
2.00	175
2.50	140
3.00	116
3.50	100
4.00	87.5
5.00	70.0
6.00	58.4
7.00	50.0

*f is the upper-bandpass frequency at the 3-dB point for a low-pass filter. The frequency f can also be considered the upper (highest) frequency component (with a significant amount of energy) in the waveform generated when a logic device switches states.

characteristics of advanced devices are two properties that the designer must understand and be aware of. Advanced TTL devices have very little noise tolerance and generate large transient currents during switching. Advanced CMOS devices have more noise tolerance than TTL but generate more noise. In either case, these two device characteristics tend to combine in a most unfavorable manner. Large transient currents generate large amounts of noise in very close proximity to devices that have very little tolerance for noise. Failing to account for the noise-generating characteristics of high-speed logic devices is certain to lead to the design of unreliable systems.

In summary, when high-speed logic devices are used, the logic design cannot be separated from the electrical and mechanical design, which has so often occurred in the past.

References

1. IEEE Standard 181-1977, "IEEE Standard on Pulse Measurement and Analysis by Objective Techniques," IEEE, 1977.
2. Southard, Robert K.: "High-Speed Signal Pathways from Board to Board," in 1981 WESCON Records, Session 18, September 1981, paper no. 2.

TTL and CMOS Logic Families

Most new designs have performance requirements that necessitate the use of high-speed advanced Schottky transistor-transistor logic (TTL), advanced complementary metal oxide semiconductor (CMOS), or emitter-coupled logic (ECL) integrated circuit (IC) logic devices. It has been recognized for some time that a great deal of attention must be given to the electrical system when ECL devices are used, but it is only recently that TTL and CMOS performance has reached a level where transmission line and other electrical system concerns associated with high-speed devices must be addressed.

To optimize the electrical environment for high-speed TTL or CMOS logic device operation, it is necessary to understand the basic characteristics of the TTL and CMOS logic families. To that end, the following discussions and tables present a brief overview of the various TTL and CMOS logic families and a comparison of the advanced logic families to the earlier, and perhaps more familiar, logic families. Certain critical interface requirements are also discussed.

2-1 TTL Logic Families

Transistor-transistor logic integrated circuit devices have been commercially available since the late 1960s. As would be expected, the features of the available TTL devices have improved considerably during the past 20 years. Compared to earlier devices, the more recently developed TTL families have higher output drive, lower input currents, and much improved speed-power products (to mention only a few of the improvements).

The major TTL logic families, preceded by their letter designations, are:

(none)	Original TTL
L	Low-power TTL
H	High-speed TTL
S	Schottky TTL
LS	Low-power Schottky TTL
F	FAST TTL
ALS	Advanced low-power Schottky TTL
AS	Advanced Schottky TTL

The logic families that are viable product lines with projected long-term availability are the ALS, the AS, and the F families. The AS and F families represent the state of the art in high-speed high-performance TTL technology.[1] The status of the older families is described below.

The original TTL family is seldom used in new designs but is frequently encountered in existing hardware. Some special medium-scale integration (MSI) parts still find some use where no equivalent part is available in the newer families. Although mainly of historical interest from a design standpoint, designers need some familiarity with the specifications of the original family since some manufacturers still reference the input-output characteristics of the newer families to the original TTL family values. For example, in FAST data books, input load levels and drive current capabilities are given in unit values that are referenced to the original TTL family.

Two of the early families, L and H, are mainly of historical interest at this time; only a small number of devices in either family are still being produced.

Of the older TTL logic families, the LS family is of the most interest to designers today. Some would class the LS family as obsolete, but the LS family still enjoys widespread use. Many designers are familiar and comfortable with the LS family and continue to use LS parts where they meet system requirements. In general, LS parts do not have sufficient speed for today's high-performance applications; they are not recommended for systems with clock frequencies greater than 8 MHz. If LS parts are used in applications where the clock frequency exceeds 8 MHz, the signal timing must be carefully evaluated. At present, LS parts are available from many sources, and as a result, they are much less expensive than AS, ALS, or FAST parts. The definition of LS and ALS parts becomes somewhat blurred when dealing with medium-scale integration and large-scale integration (LSI) parts. Many of the medium- and large-scale LS devices that are available from certain of the IC manufacturers have many of the same characteristics that are used to distinguish the ALS family from the

LS family, i.e., higher noise margin, lower input currents, and higher operating speeds than those of the basic small-scale (SSI) LS parts. How long the LS family will be available is of concern, but none of the major domestic suppliers (except perhaps those with a vested interest in ALS) seem to be planning to downgrade the availability of the more popular LS parts in the near future. Some have expressed an intent to support the LS family for a number of years.

The ALS family has been slow to establish a clear niche. Most high-performance digital systems have moved to operating frequencies that have bypassed the optimum ALS operating range. At present, ALS parts are being used in place of LS parts in many applications where long-term availability of replacement parts is of concern. It is expected that as time goes by ALS parts will replace LS parts in most applications. More sources of ALS parts are needed so that they can compete with LS on a more equal cost basis.

In the past the S family provided the means for building large high-speed TTL systems. However, the use of the S family of parts is on the decline, and its use is not recommended for new designs. Several manufacturers have discontinued production of S parts altogether, while others no longer provide the military temperature range (−55 to +125°C) parts. The FAST family, or combinations of FAST, ALS, and AS have replaced the S family for most high-speed applications.[2] The drive capabilities of S, AS, and FAST are very similar. The operating speeds of AS and FAST parts are slightly faster than those of equivalent S parts. Their edge rates (during switching) are faster than those of S, which can lead to noise problems in retrofit applications where they are substituted for S. Input load currents for AS and FAST parts are much less than those of S parts and so is the power dissipation. The higher power dissipation of S parts is a definite disadvantage in most applications. In general, AS or FAST parts can be substituted for S parts if the system has a solid electrical system. If not, the higher edge rates of AS or FAST parts may cause sufficient additional noise and signal distortion to degrade the operation of the system.

For new designs that must operate at 20 MHz or more, FAST or AS devices (or advanced CMOS devices) are required. None of the other TTL families is suitable for large systems that must operate at clock rates of 20 MHz or more. In actual system applications, there is little difference in the performance of AS or FAST parts. Certain functions in the AS family are slightly faster than equivalent FAST parts, but the speed difference is usually insignificant, and most static characteristics are similar. Input current and output drive ratings for the two families are about the same. Most SSI FAST parts dissipate slightly less power than SSI AS parts, but at the MSI level there is not much difference. If a selection must be made between the two high-speed advanced Schottky logic families, price and availability may be the best selection criteria since there is little difference in basic electrical characteristics.

2-2 CMOS Logic Families

CMOS IC logic devices have been commercially available since the early 1970s. However, it is only recently that CMOS logic devices have found widespread use. The early devices required higher than normal (for digital circuits) supply voltages, were difficult to interface to TTL devices, and were very slow. As a result, early CMOS devices found limited use. They were most often used where low power dissipation was of prime importance or in direct battery-powered applications that required devices that would operate over a wide range of supply voltages. However, the present generation of CMOS devices have overcome most of the incompatibility problems of the early families; CMOS logic devices are now available that are power-supply and signal-level compatible with TTL devices and suitable for most modern high-performance digital applications.[3] Advanced CMOS logic devices have speed and drive characteristics that are similar, and in some cases superior, to advanced Schottky TTL devices.

The major CMOS logic device categories are listed below with their letter designators when appropriate.[4]

1. Original 4000 CMOS series and updated 4000A and 4000B series devices

2. High-speed CMOS, which includes various HC and HCT devices and families of devices from numerous manufacturers

3. Advanced CMOS, which includes devices and families with letter designators such as AC, ACL, ACT, FCT, and devices with a simple C designator

Devices in each of the general categories listed above are available at this time, except that devices in the original 4000 series are no longer available and devices in the 4000A and 4000B series are approaching obsolescence. However, many of the devices in the 4000A and to a greater extent in the 4000B series are still available. There are applications, such as direct 9- or 12-V battery powered applications, where 4000A or B series parts are still the most appropriate choice.

Devices in the HC and HCT high-speed CMOS series are readily available from a large number of manufacturers. Devices in the HC series have switching thresholds at 20 and 70 percent of V_{cc}, while devices in the HCT series have thresholds levels that are compatible with TTL levels when operated with TTL supply voltage levels (i.e., 5 V ± 10 percent). High-speed CMOS HC and HCT devices are advertised as speed compatible with LS TTL devices, but in most actual applications HC and HCT devices are slower than TTL devices. High-speed CMOS devices that are pin compatible with LS-TTL and 4000 series CMOS are available. However, devices with HC designators do not have TTL-compatible input levels. Signals can be transferred from HC devices to TTL devices, but additional circuitry is required to transfer signals from TTL devices to HC devices.[5]

The advanced CMOS category encompasses a wide variety of high-speed devices and logic families. In general, most devices classified as advanced CMOS logic have speeds that are comparable to those of advanced Schottky TTL devices such as AS or FAST. Advanced CMOS devices are optimized to operate with TTL supply voltage levels, and those that are TTL-level compatible are only specified with TTL supply voltage levels of 5 V ±10 percent. Yet, most advanced CMOS devices will operate with 2- to 6-V supply levels, including those with TTL input levels, but only those with CMOS input levels are specified for such operation. Most advanced CMOS devices are pin compatible with equivalent TTL functions. Some advanced CMOS logic families have both devices where the input switching threshold is approximately centered (those with letter designators without a T) and other devices with their input thresholds shifted so that they are compatible with TTL levels (those with ACT, FCT, etc., designators). However, several advanced CMOS logic series are only offered with TTL input levels. One example is the FCT series; there is no FC series.

Of the three major CMOS logic categories, devices and logic families in the advanced CMOS category are currently of most interest to designers. Most present high-performance systems, e.g., systems that must operate with clock rates of 20 MHz or more, require advanced CMOS devices; 4000 and HC devices are too slow for most present-day applications.

The edge rates of advanced CMOS parts are faster and the transition levels greater (5 V for CMOS versus 3 V for TTL) than TTL, which can lead to serious system noise problems unless great care is taken to ensure low-impedance power, ground, and signal interconnections.[6] Most standard (i.e., not drivers) advanced CMOS devices have symmetrical ±24-mA output current drive. Thus, sink current is slightly greater than AS or FAST devices (24 versus 20 mA), and drive current is significantly greater than AS or FAST (24 versus 1 mA).

In many applications, advanced CMOS parts can be substituted inplace of AS or FAST parts, but they require a solid electrical system. If not, the higher edge rates and larger signal transitions may cause sufficient additional noise and signal distortion to degrade system operation.

2-3 Comparison of Operational Characteristics

Table 2-1 shows typical gate power dissipation as well as typical gate and clocked register speed for several TTL and CMOS logic families (which are grouped and listed in the approximate order of slowest to fastest). Typical values are shown for operation at +25°C and V_{cc} equal to 5 V (4.5 V for HC).

The TTL families fall into two broad power classifications: the S, AS, and original TTL family are in the higher power range, and the rest of the families are in a lower power range. Device dissipation has little meaning in the case of CMOS logic devices since their static device dissipation is in the microwatt range. The logic families listed, with one exception, fall into two broad speed classifications: AC, AS, F, and S are in the high-speed area, and HC, LS, and the original TTL family are in the low-speed area.[2] The ALS family is the exception; ALS devices are faster than HC or LS, yet they are slower than AC, AS, or F and slower than S in most cases. Device and system operating speed and derating considerations are treated with a great deal of emphasis and detail in Chapter 12.

It should be noted that the times shown in Table 2-1 are typical +25°C values: worst-case values at 125°C are approximately 2 times typical values. As a general rule of thumb, to convert +25°C, +5 V timing parameters, i.e., propagation delays or setup and hold times, to worst case over temperature, the following multipliers should be used:[7]

TABLE 2-1 Gate and Register—Typical Power and Speed

Family	Power, mW	Gate speed,* ns		Register speed,* ns	
		15-pF load	50-pF load	t_{prop}	t_{setup}
(TTL)	10	9	—	27	13
LS	2	9.5	13	23	17
ALS	1	4	9	13	10
S	19	3	5	8	2
F	4	2	3.5	6.2	2.6
AS	10	1.5	3.5	7	3
HC	†	—	18	35	20
AC	†	—	6	6	1

*Typical values at +25°C and 5 V (4.5 V for HC) are shown.
†Power of AC and HC is a function of frequency and load.

- To convert typical +25°C timing parameters to worst case over the commercial temperature range, multiply typical +25°C timing parameters by a factor of 1.5.

- To convert typical +25°C timing parameters to worst case over the military temperature range, multiply typical +25°C timing parameters by a factor of 2.

- To convert maximum (or minimum) +25°C timing parameters to worst case over the commercial temperature range, multiply maximum (or minimum) +25°C timing parameters by a factor of 1.25.

- To convert maximum (or minimum) +25°C timing parameters to worst case over the military temperature range, multiply maximum (or minimum) +25°C timing parameters by a factor of 1.5.

2-4 Static Input-Output Characteristics

The following group of tables list dc current drive and voltage level characteristics for several of the more common TTL and CMOS logic families. They are not intended to be studied in great detail but are provided as a quick reference source. They are based on a standard NAND gate (a 54/74 __ 00-type gate). In general, the drive and load levels shown are representative of all standard devices in a family, but caution must be exercised because some devices within a family may have different characteristics, and *drivers,* by definition, have more drive than is listed. The device specifications for a given device should always be consulted and closely studied before the device is applied. Failure to read and understand the device specifications or data sheets remains one of the major sources of design error. The specifications or data sheets should be read over and over. A great deal of attention should be given to the notes and fine print. For complex devices, it is best to find and consult someone that has some experience with the part of interest; however, when the part does not work as expected, the fact that you talked to someone about it is not considered a very good excuse for not having studied the data sheet.

Table 2-2 shows the current drive capability and the input load current specifications for standard parts (drivers have higher drive ratings) for the logic families of most interest.

Table 2-3 shows the worst-case-specified input- and output-voltage levels for the families of interest. The difference between the worst-case input and output levels is the tolerance that a given set of de-

TABLE 2-2   Input Load–Output Drive for the 74/54 __ 00 NAND Gate

Family	Input current*		Output current*	
	I_{IL}, mA	I_{IH}, µA	I_{OL}, mA	I_{OH}, mA
(TTL)	−1.6†	40	16	−0.4
LS	−0.4	20	4	−0.4
ALS	−0.1	20	8	−0.4
S	−2.0	50	20	−1.0
F	−0.6	20	20	−1.0
AS	−0.5	20	20	−2.0
HCT	±0.001	±1	4	−4
HC	±0.001	±1	0.02	−0.02
ACT	±0.001	±1	24	−24
AC	±0.001	±1	24	−24

*Current out of a terminal is given a negative value.
†Note the −1.6 mA *low* input-current and 40 µA *high* input-current specification for standard TTL parts. The unit load and drive ratings given in data books for some of the newer logic families are referenced to the original TTL values of −1.6 mA and 40 µA. Also, note that the S family has the highest input loading, and the HC and LS families the lowest output drive capability.

vices, or a given system built with a particular set of devices, has for imperfection in the reference system and in the interconnection system. This tolerance is known as noise margin and is discussed in some detail in Chapter 4.

Table 2-4 shows the input- and output-voltage levels at which the currents shown in Table 2-2 are specified. Note that the voltage levels at which the input or output currents are specified differ slightly in some cases from the worst-case-specified input- or output-voltage levels (see Table 2-3).

TABLE 2-3   Specified Worst-Case Input-Output Voltage Levels for the 74/54 __ 00 NAND Gate

Family	Input levels		Output levels	
	V_{IL}(max), V	V_{IH}(min), V	V_{OL}(max), V	V_{OH}(min), V
(TTL)	0.8	2.0	0.4	2.4
LS	0.7	2.0	0.4	2.5
ALS	0.8	2.0	0.5	2.5
S	0.8	2.0	0.5	2.5
F	0.8	2.0	0.5	2.5
AS	0.8	2.0	0.5	2.5
HCT	0.8	2.0	0.4	3.7
HC	0.9	3.15	0.1	4.4
ACT	0.8	2.0	0.4	3.7
AC	1.35	3.15	0.1	4.4

TABLE 2-4   Specified Input-Output Current* and Voltage Levels for the 74/54 _ 00 NAND Gate

Family	I_{IL}, mA at V_{IL}, V	I_{IH}, μA at V_{IH}, V	I_{OL}, mA at V_{OL}, V	I_{OH}, mA at V_{OH}, V
(TTL)	1.6,0.4	40,2.4	16,0.4	0.4,2.4
LS	0.4,0.4	20,2.7	4,0.4	0.4,2.5
ALS	0.1,0.5	20,2.7	8,0.5	0.4,2.5
S	2.0,0.5	50,2.7	20,0.5	1.0,2.5
F	0.6,0.5	20,2.7	20,0.5	1.0,2.5
AS	0.5,0.5	20,2.7	20,0.5	2.0,2.5
HCT	0.001,0	1,V_{cc}	4,0.4	4,3.7
HC	0.001,0	1,V_{cc}	0.02,0.1	0.02,4.4
ACT	0.001,0	1,V_{cc}	24,0.4	24,3.7
AC	0.001,0	1,V_{cc}	24,0.4	24,4.4

*Current magnitude is shown; see Table 2-2 for sign.

THE VALUES SHOWN IN TABLES 2-2, 2-3, AND 2-4 ARE APPLI- CABLE TO MOST STANDARD PARTS (NOT DRIVERS) IN THE RESPECTIVE FAMILIES FOR BOTH 0 TO +70°C COMMERCIAL 74 _ XX TEMPERATURE RANGE PARTS AND FOR -55 TO +125°C MILITARY 54 _ XX PARTS, BUT IF THE RATING IS DIF- FERENT, THE -55 TO +125°C MILITARY RATING IS LISTED.

Unit loads.   Many TTL device data sheets list drive and load capabil- ities or requirements as a function of a standardized unit load rather listing the actual current. A unit load is usually defined as the load for one input of a standard gate for that family of devices. However, some data books list load and drive capability in unit loads that are based on the original TTL family, i.e., a unit load equals 1.6 mA for a *low* input and 40 μA for a *high* input. Table 2-5 lists the same data as Ta- ble 2-4, but in unit load form (Table 2-5 only includes data for the TTL families). In Table 2-5 a unit load is referenced to a given family and

TABLE 2-5   TTL Input Load–Output Drive for the 74/54 ____ 00 TTL NAND Gate

Family	In unit loads			
	Input load		Output drive	
	I_{IL}	I_{IH}	I_{OL}	I_{OH}
(TTL)	1	1	10	10
LS	1	1	10	20
ALS	1	1	80	20
F	1	1	33.3	50
S	1	1	10	20
AS	1	1	40	100

not back to the original standard TTL family. That is, LS is referenced to LS and F is referenced to F. Note also that *high* and *low* unit loads are different. It must be emphasized that the data shown in Table 2-5 is only good for standard SSI-type parts such as gates; drivers and other special purpose parts may have different ratings.

It is not a common practice to list CMOS input loads or drive capability in terms of unit loads. Since CMOS devices have very high input impedances and very low input currents, unit dc load or drive has very little meaning and as a result is seldom used in reference to CMOS devices.

2-5 Compatibility of Logic Families

In all cases where logic families are mixed, dynamic as well as static compatibility must be considered. However, the following discussion only addresses static interface compatibility.

The various TTL families and CMOS families with TTL input levels are voltage level and current drive compatible and can be mixed with great care. Their input-output voltage levels are not exactly the same, but they are within an acceptable range. The slight variation in levels cause some reduction in *low* noise margin when other Schottky TTL devices are used to drive LS devices (100 mV), and *high* noise margin is enhanced when CMOS devices are used to drive TTL devices. Nonetheless, as long as current drive specifications are not exceeded, TTL- and TTL-input-compatible CMOS (HCT and ACT) families can be mixed. Current drive capability of TTL-input-compatible CMOS families are similar to that of comparable TTL families, and CMOS input currents are of such a low magnitude that all TTL families can source or sink several orders of magnitude more dc current than is needed to drive a very large number of CMOS inputs. Extra care must be taken to ensure that signals are not overloaded in systems with mixed logic families, particularly in those systems where devices with high input-current requirements are mixed with low drive HC or LS devices. In all applications, an audit of signal loading should be performed as one of the final design tasks.

The various TTL families and CMOS families with CMOS input levels cannot be mixed without the use of additional interface circuitry for TTL-to-CMOS signal transfers. The output-voltage levels of CMOS devices are compatible with TTL input levels, but CMOS input levels are not compatible with TTL output levels. Thus, additional circuitry is required to transfer signals from TTL to CMOS voltage levels, but it is not for the other direction (see Section 2-6).

Table 2-6 shows the current drive capability of one logic family versus another. However, the number of CMOS loads that TTL devices or

TABLE 2-6   DC Drive Cross-Reference for TTL-Level-Compatible Families

DC Drive Capability* (no derating, standard 54 ____ XX, −55°C to +125°C devices)

Driving device						
	Number of loads that can be driven					
Family	LS	(TTL)	S	F	AS	ALS
(TTL)	40	10	8	26.7	32	20
LS	10	2.5	2	6.7	8	20
ALS	20	5	4	13.3	16	20
S	50	12.5	10	33.3	40	50
F	50	12.5	10	33.3	40	50
AS	50	12.5	10	33.3	40	100
HCT	10	2.5	2	6.7	8	40
ACT	60	15	12	40	48	240

*Low input current I_{IL} determines the maximum number of devices that can be driven for all the families, except when the ALS family is driven by other TTL devices. When ALS is driven with other TTL devices, the $high$ input current I_{IH} determines the maximum number of devices that can be driven (see Table 2-2).

other CMOS devices can drive are not cross referenced since CMOS dc input currents are of such a low magnitude that such information is meaningless. Manufacturers' maximum ratings are shown with no design derating allowances.

2-6   TTL-to-CMOS and CMOS-to-TTL Interface Requirements

Great care is required where signals must cross TTL-CMOS interfaces. In general, TTL output voltage levels are not compatible with CMOS input-voltage level requirements. However, there are some exceptions to the above; most of the newer CMOS logic families include some devices that are designed to be compatible with TTL levels. Those CMOS devices with TTL-compatible input levels interface directly with TTL devices and require no additional interface circuitry.

The CMOS logic devices that have input levels that are compatible with TTL levels have logic family letter designators that end with a T. For example the ACT, FCT, and HCT families have input thresholds that are designed to be compatible with TTL levels. No additional circuitry is required to transfer signals from TTL sources to ACT, FCT, or HCT devices (see Figure 2-1).

Devices with "normal" or "true" CMOS input levels are not compatible with TTL output levels, and additional circuitry is required to transfer signals from TTL to CMOS levels. The CMOS logic devices that have CMOS input levels have family letter designators such as AC or HC. Devices with CMOS input levels are TTL low-level compatible but are not TTL $high$-level compatible. The minimum $high$-

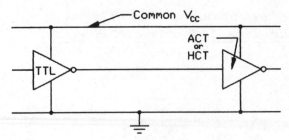

Figure 2-1   TTL output to ACT, FCT, or HCT inputs.

output specification of TTL devices is 2.4 or 2.5 V, and the minimum *high* input requirement for CMOS devices is 3.15 V (see Table 2-3). Thus, when signals must be transferred from TTL devices to CMOS devices with CMOS thresholds, level shifters or pull-up resistors are required (see Figure 2-2). A number of problems are associated with pull-ups, such as degraded rise times, extra power dissipation, and additional current sinking requirements for drivers. The best means of transferring from TTL levels to CMOS level is to use ACT, FCT, HCT, etc., CMOS devices with TTL input levels and CMOS output levels.

Advanced and HC CMOS logic devices have output levels that are compatible with TTL input levels when powered from a common source. *Low* CMOS output levels are approximately the same as TTL output *low* levels. *High* CMOS output levels are greater than normal TTL *high* levels, but do not exceed maximum TTL input limits. In many applications, higher *high* levels are of benefit. In noisy applications, they provide more noise margin. Higher *high* levels do have some disadvantages. They generate more system noise because of the larger signal transitions, and increase *high*-to-*low* signal response because they require more time to reach *low* switching thresholds. However, in most applications, transferring signals from CMOS devices to TTL devices requires no special circuitry provided that all devices have the same power supply (see Figure 2-3).

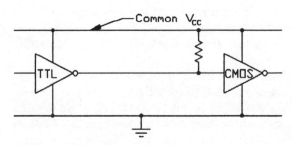

Figure 2-2   TTL output to AC or HC inputs.

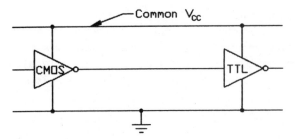

Figure 2-3   CMOS output to TTL input.

In general, 4000 series CMOS devices are not TTL level or current compatible. In most applications, 4000 series devices are powered with supply voltage levels much in excess of 5 V, and thus *high* input and output levels are not compatible with TTL *high* levels. Special level-shifting circuits are required to translate levels.

Failure to provide proper levels to CMOS inputs may result in destruction of parts as well as logic malfunctions. Input levels that fall between logic *high-* and *low-*input limits and inputs that exceed V_{cc} or ground are both classified as improper inputs and both can be destructive. In general, CMOS parts with "true" CMOS inputs require inputs to have levels between ground and 30 percent of V_{cc} or between 70 percent of V_{cc} and V_{cc} to ensure that one or the other of the field-effect transistors (FETs) in the input complementary structure are OFF. When input levels fall between 30 and 70 percent of V_{cc}, both FETs in the input complementary structure may turn ON and current flow from V_{cc} to ground through the input complementary structure can result in overheating of parts and possible eventual failure. Such a condition exists when *high* TTL levels are used to drive CMOS-level inputs. Worst-case TTL minimum *high* output levels, which are 2.4 or 2.5 V, do not fall within the range between V_{cc} and 70 percent of V_{cc}. Hence, pull-ups or other kinds of circuitry are required on TTL-to-CMOS interfaces to ensure that inputs do not dwell in the danger zone.

When input levels exceed V_{cc} or go below ground, even for a very short time or by a very small amount, CMOS devices can be damaged. When inputs or outputs exceed V_{cc} or ground, parasitic SCRs in electrostatic protection circuits can be triggered ON, which can result in excessive V_{cc} to ground currents and destruction of parts. If input (or output) levels can exceed V_{cc} or ground when devices are powered, some means of current limiting must be provided so that the current levels needed for SCR triggering are not reached (see Section 3-2-6). It is also important to limit current at interfaces where one side of the interface may be powered before the other side is. If current is allowed to flow into input (or output) protection networks of unpowered devices, those devices may latch-up when power is applied. It is generally advisable to have some resistance in series with CMOS circuit board or system interfaces where low-impedance sources (or loads) can be encountered in order to ensure that input- or output-current ratings are not exceeded.

2-7 Derating of Manufacturers' Drive Specifications

In critical applications, manufacturers' output drive specifications should be derated. The input-output voltage levels and current load and drive ratings shown on most logic device data sheets are defined and tested under static conditions. It cannot be assumed that static ratings are adequate for all dynamic situations. If a part is not tested in a configuration exactly as it is used, there is always some uncertainty as to how it may function. Drive derating is one method used to compensate for that uncertainty. Drive derating also helps compensate for device degradation with aging and worst-case combinations of electrical and environmental stress.

In benign applications, not exceeding the manufacturer's ratings may be adequate to ensure a reliable design. However, in high-performance applications, it is best if devices are not fully loaded, which leaves some margin for uncertainties. In critical applications, such as military or space applications, an extra allowance is needed to ensure that parts will function at the extremes of the operating voltage, frequency, and temperature range and in the face of long-term aging and other degenerative effects. The typical derating criteria used for systems that must operate over the full military temperature range of −55 to +125°C is to limit *fan-out*, or actual load current of devices, to no more than 70 percent of the manufacturer's specified value. For space applications a derating factor of 50 percent is often used.

Derating of manufacturers' specifications should not be limited to dc considerations. In all applications, ac load parameters must also be considered. When the newer high-speed advanced Schottky TTL or advanced CMOS logic families are used in high-speed systems, ac drive limits usually restrict system operation rather than dc drive considerations (see Chapter 12).

2-8 Special Drive and Load Considerations

2-8-1 HC or LS driving S devices

Static (dc) drive limits are often inadvertently exceeded in applications where LS or HC parts are used to drive S devices, or others, such as standard (original) TTL parts or FAST octal buffers, that have high input-current requirements. Various combinations of S and LS parts that can be driven by an LS (or HC) device with 4-mA current sink capability are shown below.

Caution—LS or HC driving S often leads to drive problems.

How many S and LS parts can be driven with an LS (or HC) part?

1. No derating—LS drive is specified as 4 mA.

$$2 \text{ S unit loads} = 4.0 \text{ mA}$$

or

$$
\begin{array}{ll}
1 \text{ S unit load} & = 2.0 \text{ mA} \\
\text{and } 5 \text{ LS unit loads} & = \underline{2.0 \text{ mA}} \\
& 4.0 \text{ mA}
\end{array}
$$

2. 30% derating—4 mA × 70% = 2.8 mA.

$$
\begin{array}{ll}
1 \text{ S unit load} & = 2.0 \text{ mA} \\
\text{and } 2 \text{ LS unit loads} & = \underline{0.8 \text{ mA}} \\
& 2.8 \text{ mA}
\end{array}
$$

3. 20% derating—4 mA × 80% = 3.2 mA.

$$
\begin{array}{ll}
1 \text{ S unit load} & = 2.0 \\
\text{and } 3 \text{ LS unit loads} & = \underline{1.2 \text{ mA}} \\
& 3.2 \text{ mA}
\end{array}
$$

2-8-2   Special load requirements

Another drive requirement that is often overlooked is that some inputs on some TTL devices, particularly on the older TTL families, have more than a one unit load requirement. Some examples are listed below.

Caution—Some older TTL devices have more than a one unit load input.

Devices with More Than One Unit Input Load

Device	Input	Number of unit loads
5474	CLEAR & CLOCK	2
54109	SET & CLOCK	2
	CLEAR	4
5485	DATA	3
74/54161	CLOCK & CLOCK ENABLE	2

In addition, in a very limited number of cases, the same generic part from different manufacturers has different input loading (the 5483 is such a part). Fortunately, most of the newer logic families have buffered inputs so that inputs present only one unit load to the external world. The designer should always check the most authoritative parts specification available to ensure that the proper load value is known and factored into the design.

2-8-3 Loading when inputs common to a given TTL gate are tied together

Sometimes unused inputs on a given gate (NAND, NOR, etc.) are tied to a used input on the same gate as a convenient means of ensuring that the unused input is always maintained at a valid logic level (inputs should never be left floating; see Chapter 14). For a *low* input, tying inputs of a standard TTL gate together does not increase the *low* load current I_{IL} that the driver must sink. That is, for a given gate, regardless of the number of inputs it may have, if they are tied together, they present only one load for the *low* input state. However, the total *high*-input-load current is increased by the number of inputs tied in parallel. Fortunately, most of the families have more drive, i.e., they can drive more unit loads in the *high* state than they can in the *low* state (see Table 2-5). Only the original TTL family and the ALS family are specified with equal or less unit load drive in the *high* state than in the *low* state. The remainder of the families that are of interest have at least twice the drive in the *high* state as in the *low* state. The reason for the different loading for *high* and *low* states when inputs are tied together is described in Chapter 3.

The dynamic (ac) loading is increased when inputs are paralleled; thus such an arrangement is not recommended when speed is of concern. Specifically, paralleling inputs on LS gates is not recommended; such an arrangement reduces the noise tolerance of the used input. Paralleling LS inputs parallels the capacitance of the input diodes and increases the capacitance between the signal and the interior of the device. The increased capacitance increases the possibility of noise being coupled into and upsetting internal nodes.

2-9   Increasing Drive Capability

How can drive or fan-out capability be increased when additional drive is needed? The most obvious technique is to split a heavily loaded signal into segments and use multiple devices to drive the separate segments. The next most obvious technique is to use drivers, such as 54/74 __ 37 or the 54/74 __ 240 series of octal drivers, to buffer outputs from low drive devices. When subdividing the load or adding drivers does not alleviate the problem or is not practical, drivers can be paralleled.

Outputs of TTL devices on the same chip or driver in the same package can be paralleled to increase fan-out capability when there is no other alternative for achieving the needed drive. When paralleling TTL drivers, it is important to do so only with drivers in a common package so that the drivers share the load. Drivers in a common package can be expected to see approximately the same environmental conditions. They should see the same temperature and supply voltage. They should also have similar device characteristics as a result of common processing. Thus, TTL drivers in a common package tend to share the load, whereas devices in different packages may not. Devices in different packages may have slightly different characteristics as a result of different processing and environments, and thus be less likely to share the load. If the devices do not share the load, one device may be over stressed as a result of having to handle a large portion of the load (such a condition is sometimes referred to as "load hogging").

Paralleling CMOS devices to achieve more current drive capability is an acceptable practice. Enhancement-mode field-effect transistors, as used in CMOS devices, have a positive ON impedance temperature coefficient; as a result paralleled CMOS outputs are self-regulating to some degree. For example, if a large portion of the load current is flowing in one of several paralleled devices, that device tends to heat up more than the other paralleled devices; as a result its resistance goes up and its portion of the current goes down. Thus, paralleled CMOS devices tend to share the load. Still it is best to only parallel CMOS devices in a common package.

Methods of increasing (drive) fan-out

- Separate the load and use multiple drivers
- Use high current drivers (74/54 __ 37, 74/54 __ 240, 244, etc.)
- Parallel outputs of devices in the same package

References

1. Hall, Bill: "Beware of Sunset Logic Technologies," *Electronic Products,* October 15, 1987, pp. 47–52.
2. Greer, W. T., Jr.: "Advanced Schottky Logic, Gearing up for High Performance," *Electronic Products,* July 2, 1984, pp. 49–54.
3. Walsh, M. J.: *Choosing and Using CMOS,* McGraw-Hill, New York, 1986.
4. Frederiksen, Thomas M.: *Intuitive IC CMOS Evolution,* National Semiconductor Corp., Santa Clara, Calif., 1984.
5. Funk, Dick: "Design Guidelines for CMOS Logic Systems," *Electronic Products,* March 28, 1984, pp. 75–79.
6. *Advanced CMOS Logic Designer's Handbook,* Texas Instruments Inc., Richardson, Tex., 1987.
7. *Bipolar Microprocessor Logic and Interface Data Book,* Advanced Micro Devices Inc., Sunnyvale, Calif., 1981.

Bibliography

Advanced CMOS Logic Data Book, Texas Instruments Inc., Dallas, Tex., 1988.
ALS/AS Logic Data Book, Texas Instruments Inc., Dallas, Tex., 1986.
CMOS Logic Databook, National Semiconductor Corp., Santa Clara, Calif., 1988.
FACT—Advanced CMOS Logic Databook, National Semiconductor Corp., Santa Clara, Calif., 1988.
FAST—Advanced Schottky TTL Databook, National Semiconductor Corp., Santa Clara, Calif., 1988.
High-Speed CMOS Logic Data Book, Texas Instruments Inc., Dallas, Tex., 1986.
TTL Logic Data Book, Texas Instruments Inc., Dallas, Tex., 1988.

TTL and CMOS Circuits

Experience has shown that high-speed digital devices cannot be applied successfully when the design effort is limited to functional logic considerations only. The electrical characteristics and the limitations of the devices being applied must be understood. A thorough understanding of the operation and the limitations of the interface circuitry is of prime importance.

3-1 TTL Circuits

Most early TTL data sheets included schematics, but that is no longer the case. Current data sheets seldom show internal circuitry; thus many digital designers have had no exposure to TTL circuitry. A knowledge of the minute details of the internal structure of TTL devices is not necessary for system designers, but an understanding of basic TTL circuit operation and the critical parameters associated with the input-output circuitry is important. It must be emphasized that the power and ground pins are two of the most critical interfaces. If problems that require some knowledge of device internal circuit operation are encountered, schematics of basic small-scale IC (SSI) devices, such as gates and drivers (see Figure 3-1), can sometimes be found in the introductory section of data books. However, for medium-scale IC (MSI) devices, such as 4-bit counters, and for large-scale IC (LSI) devices, such as ALUs, the most that is provided is a functional-block-level representation. It would be impractical to show a detailed internal schematic, and it would be of little use to the system designer.

3-1-1 Equivalent circuits for diodes and transistors

Analyzing and understanding TTL circuit operation is greatly simplified by substituting simple static equivalent circuits or models for the

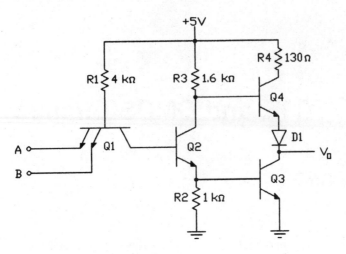

Figure 3-1   54/7400 NAND.

internal device diodes and transistors. The diode and transistor equivalent circuits, or models, shown and described below are not exact, but they are adequate for most static analysis.

Diode equivalent circuits.   Transistor-transistor logic circuits use both standard silicon and metal-silicon diodes. Standard silicon diodes are formed by p- and n-type semiconductor material junctions as shown in Figure 3-2a.[1] Schottky or metal-silicon diodes are formed at metal-silicon junctions as shown in Figure 3-2b.[2] The schematic, or circuit diagram symbol, for a standard silicon diode is shown in Figure 3-3a and that for a Schottky or metal-silicon junction diode in Figure 3-3b. When diodes are forward-biased, that is, when current is flowing in the direction shown in Figure 3-3, a battery or dc-voltage-source symbol, along with a resistor, is a useful static equivalent circuit[3] (see Figure 3-4). In general, when dealing with TTL circuits, the resistor can be neglected since, at the current levels of TTL circuits, its effect is usually negligible. Typical diode IV curves for silicon diodes and for metal-silicon diodes are shown in Figure 3-5a and b.[4] As can be seen in Figure 3-5, diodes do not have a sharp cut-on or cut-off IV relationship as is implied by a fixed-battery equivalent circuit, but for most

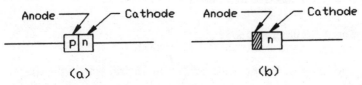

Figure 3-2   Semiconductor diode.

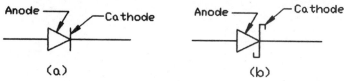

Figure 3-3   Diode symbols.

practical purposes a fixed voltage source is an adequate representation of a forward-biased diode. Diodes in TTL circuits usually operate in the flat portion of the *IV* curve. The forward-voltage drop at room temperature (+25°C) flattens out near 0.8 V for silicon diodes and near 0.3 V for Schottky diodes (see Figure 3-5). The exact value for any given diode depends upon a number of conditions, such as temperature, current flow, and basic device parameters. For the purpose of understanding TTL device operation, the exact value is not of great importance. It is not critical whether 0.7, 0.8, or 0.9 V is used for silicon diodes, or 0.2, 0.3, or 0.4 V for Schottky diodes.

For forward currents within the designed operating current range, the diode forward-voltage drop does not change much with changing current levels, but forward drop does change as a function of temperature. The forward-voltage drop across both silicon and Schottky diodes decreases by about 2 mV per centigrade degree increase in temperature (for currents in the normal operating range).[5]

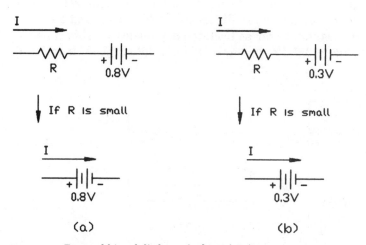

Figure 3-4   Forward-biased diode equivalent circuits.

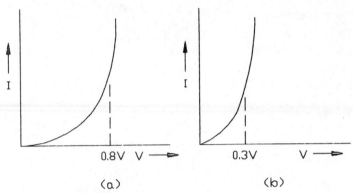

(a)

(b)

Figure 3-5   Diode *VI* curves.

Remember—Silicon diode forward-voltage temperature coefficient is −2 mV/°C.

In most TTL circuit applications, when diodes are reverse-biased as shown in Figure 3-6*a*, it is safe to treat them as open circuits with no current flow. However, reverse-biased diodes are more correctly modeled as current sources as shown in Figure 3-6*b*, where *I* is the leakage current. All diodes have some amount of leakage; however, for a simplified functional analysis of digital circuits, it is safe to assume an open circuit.

Transistor equivalent circuits. Some common graphical representations of transistors are shown in Figure 3-7.[1,3] Note that under certain conditions transistors appear as back-to-back diodes, as shown on the right side of Figure 3-7. For example, when isolated "good" transistors

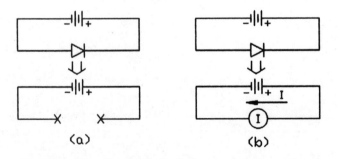

Figure 3-6   Reverse-biased diode equivalent circuit.

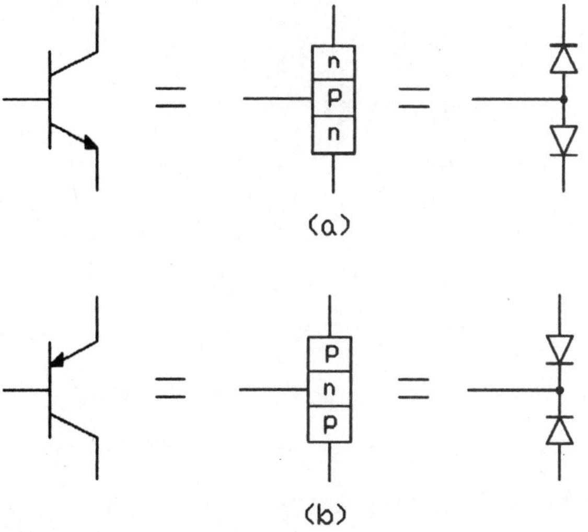

(a)

(b)

Figure 3-7 Transistor symbols, semiconductor properties, and back-to-back diode equivalent circuits.

are checked for continuity between base and collector and between base and emitter, they look like a pair of back-to-back diodes. It is important to keep the back-to-back diode analogy of transistors in mind when analyzing TTL circuits.

When dealing with digital circuits, the linear operation of transistors is of little concern. Digital designers only have to deal with transistors in two states, ON or OFF. Figure 3-8 shows the equivalent circuit for an ON and an OFF *npn* transistor, and Figure 3-9 shows the same for an ON and an OFF *pnp* transistor. The values shown for the battery or dc voltage sources are typical numbers. Whether 0.6 or 1.0 V is used for the base-emitter drop will not make a great deal of difference. Likewise, whether 0.2, 0.4, or 0.5 V is used for the ON value of collector-to-emitter drops is of no great significance.

The Schottky families of TTL devices use Schottky diode clamps connected between the base and the collector of most transistors to speed up turn OFF. Schottky diode clamps keep clamped transistors from going into saturation by diverting current from the base.[3] A nonsaturated transistor turns OFF faster than a saturated one.[6] Schottky diode clamps greatly enhanced the viability of TTL technology, and they are the key to a high-speed low-power, economical manufacturing process.

An ON Schottky *npn* transistor and its equivalent circuit are shown in Figure 3-10. The OFF condition is the same as for a conventional silicon transistor and is shown in Figure 3-8*b*.

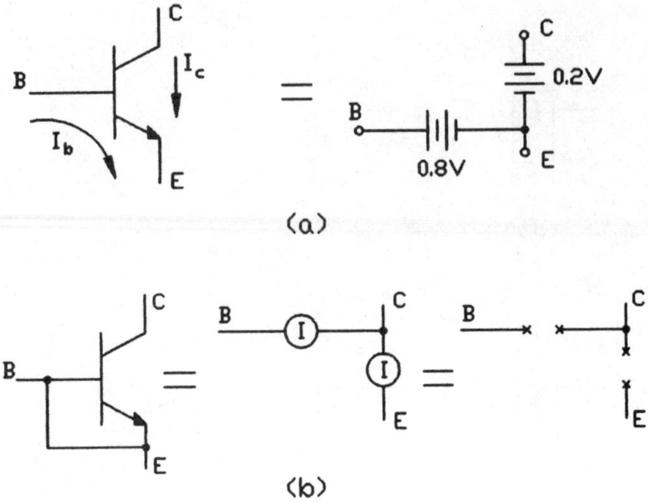

(a)

(b)

Figure 3-8 Equivalent circuits for *npn* transistors.

3-1-2 TTL circuit operation

An examination of the operation of a basic 54/7400 two-input NAND gate (see Figure 3-1) is useful for insight into all TTL circuit operation.[6,7] A two-input 54/7400 NAND gate is a simple device by today's standards. Nonetheless, the operation of the input, output,

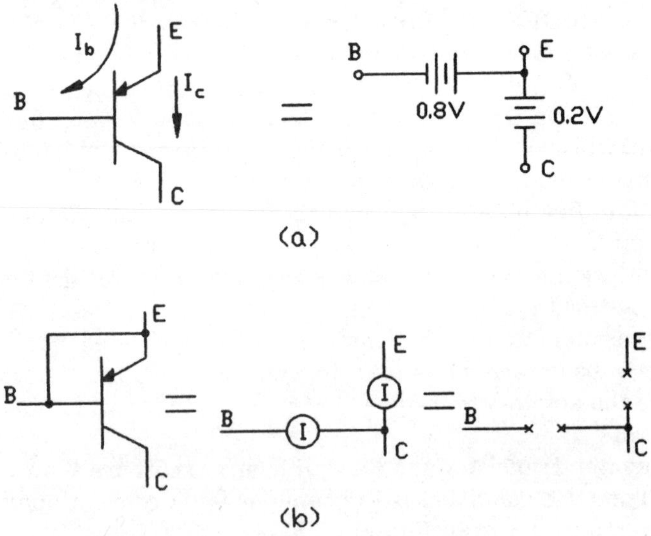

(a)

(b)

Figure 3-9 Equivalent circuits for *pnp* transistors.

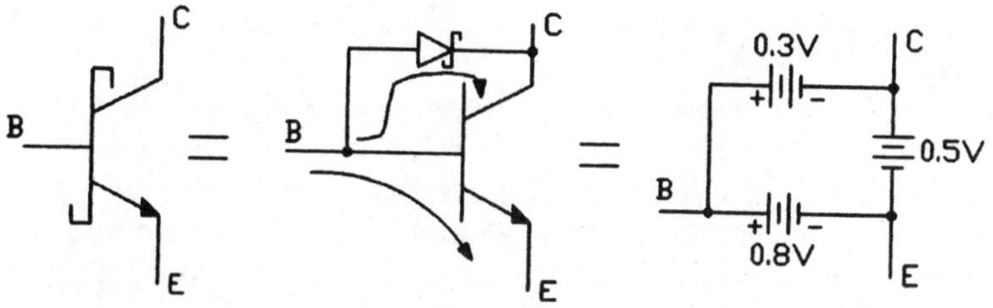

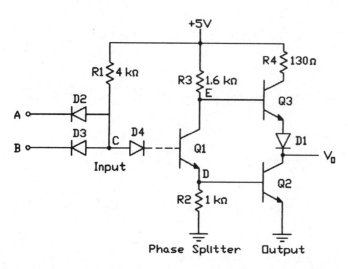

Figure 3-10 Schottky-clamped *npn* transistor.

and phase-splitter circuits used to implement basic 54/7400 devices remains fundamental to all TTL circuit operation. More complex advanced Schottky devices are structured around the same basic circuitry,[8] but as would be expected, the newer devices have many circuit enhancements which lower input currents, provide more drive, etc. However, when stripped of the enhanced circuitry, an advanced Schottky NAND gate appears and operates much the same as the circuit shown in Figure 3-1.

The following analysis and results are not exact but are useful approximations.

Operation of a two-input TTL NAND.   To simplify the initial functional analysis of the two-input NAND shown in Figure 3-1, the multiple-emitter input structure is replaced with a simple equivalent diode structure (see Figure 3-11). The change does not affect the basic functioning of the gate, but removes the somewhat more complex multiple-

Figure 3-11 Basic two-input NAND gate.

emitter structure for the moment. The operation of multiple-emitter inputs is treated in Section 3-3.

The basic gate is composed of three functional areas (see Figure 3-11): an input section, a phase-splitter section, and an output section. Most TTL devices have similar functional areas. More complex functions have additional circuitry, but the same input, output, and phase-splitter circuitry is used. For example, a register or latch must contain some additional circuitry to store information.

Equivalent circuit with a *low* input.   Figure 3-12 shows the basic gate (Figure 3-11) with diode and transistor equivalent circuits substituted for normal diode and transistor schematic representations. When an input B is grounded as shown in Figure 3-12, the associated diode is forward-biased and current flows out of the device and into ground. Currents flowing out of TTL devices are defined as negative. Thus, *low* input currents, $I_{in\ low}$ or I_{IL}, and *high* output currents, $I_{out\ high}$ or I_{OH}, are given negative signs. It is, perhaps, awkward to think of a current coming out of a device as negative, but the sign given to current flow is of little functional consequence. For ac devices, the direction of the current flow changes each half cycle. Input-output current flow for TTL devices is somewhat analogous; it changes direction with the state of the signal. If a signal is *low,* current flows out of the driven devices and into the source. If it is *high,* it flows out of the source and into the load.

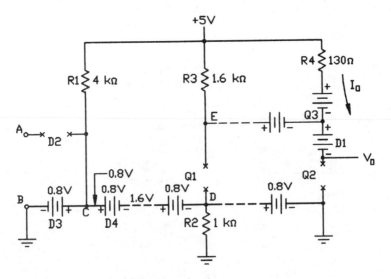

Figure 3-12   Equivalent circuit for a *low* input.

Low current I_{IL} is calculated using the portion of the gate equivalent circuit shown in Figure 3-13 and in Equation (3-1).

$$|I_{IL}| = \frac{V_{cc} - V_f}{R_1}$$

(3-1)

$$= \frac{5\text{ V} - 0.8\text{ V}}{4\text{ k}\Omega} = 1.05\text{ mA}$$

The different logic families have different values of resistors in the input section. Schematics for the original standard TTL family typically show 4 kΩ, as used in the example. The input resistor value is much higher in the lower power families; it is typically 20 kΩ for the LS family. However, regardless of the value of the input section resistor, the same general method can be used to calculate the typical *low* input current for the other families. The calculation will not result in the exact input-current values specified in a manufacturer's data books, and exact values should not be expected. The manufacturer's device specification must allow for all worst-case conditions. The models produce typical values since typical parameter values are used. If worst-case internal component parameters are used in the models, the models will yield worst-case currents. Some might ask, "Why bother looking at input currents, etc., if the derived parameter values are not exact numbers?" One reason is to understand the nature of the current flow. If the operation of a circuit is understood, special or unusual conditions can be better addressed.

The next step in the overall gate analysis is to determine whether Q1 is turned ON or OFF. If Q1 is ON, it must have base-emitter current flow. For Q1 to have base current, the voltage at node C must be greater than 1.6 V (the base-emitter drop of Q1 plus the forward drop across D4), but the voltage is only 0.8 V at node C with an input grounded; thus Q1 must be OFF. It does not matter which input is grounded, or if both are grounded, the voltage at node C is 0.8 V

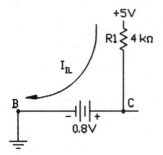

Figure 3-13   Input with a grounded input.

for either case. Two 0.8 V voltage sources in parallel or one 0.8 V voltage source between node C and ground results in the same voltage at node C. With $Q1$ in a nonconducting, or OFF, state, its collector-to-emitter, nodes E to D, are drawn open. The next step is to determine whether $Q2$ is ON or OFF. Since node D is at 0 V because $Q1$ is OFF, there is no source of base current for $Q2$; hence, $Q2$ is OFF also. Such a result should be expected since the device is a NAND gate, and a NAND with a *low* input should have a *high* output. However, at this point, it has only been established that the circuit does not have a *low* output. It has not been shown that the circuit has a *high* output; that is, $Q3$ is ON.

To show that $Q3$ is ON and that the circuit has a *high* output, it must be shown that $Q3$ has base-emitter current flow. From inspection of Figure 3-12 it can be seen that $Q3$ has a source of base current from node E as long as the output voltage V_o is less than two diode drops below V_{cc}. The two diode drops are those of $D3$ and the base-emitter (junction) drop of $Q3$. The output pull-up circuitry is arranged so that $Q3$ has a source of base current to provide the drive to quickly transition the output to a *high* level. Once an output reaches a *high* level, output current is no longer needed when driving normal TTL loads. Power is saved by reducing the drive once a static condition is achieved. However, it should be noted that output *high* levels, under all circumstances, are a function of load.

Output *high*-voltage levels.   The above analysis of the basic gate with a *low* input established that a *low* input on any or all of the inputs turns $Q3$ ON and pulls the output V_o *high*. The magnitude of V_o *high* depends on the value of the load and the output current. There are two sources of output *high* current: I_c through the collector-emitter of $Q3$ and I_b through the base-emitter of $Q3$ as shown in Figure 3-14. The current I_b, which must exist for $Q3$ to be ON, establishes one limiting value for the output voltage V_o which is

$$V_o = V_{cc} - V_{BE}(Q3) - V_f(D1) - I_b(R3) \qquad (3.2)$$

However, the load resistance R_L must be known to determine I_b. Using Equation (3-2) for the example circuit, V_o as a function of I_b is

$$V_o = 5\text{ V} - 0.8\text{ V} - 0.8\text{ V} - (I_b)(1.6\text{ k}\Omega)$$

Notice that the upper limit for a *high* output is 3.4 V (5 V − 1.6 V) when $I_b = 0$ (infinite load impedance).

In high-current situations, the collector-emitter current path I_c *through* $Q3$ may establish the output *high* level rather than the cur-

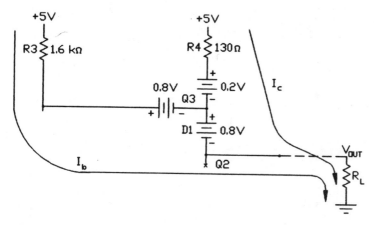

Figure 3-14 The *high*-output equivalent circuit.

rent path through the base I_b. The conditions of both paths must be calculated to establish which path is dominant. To determine V_o in the collector path limited case, calculate the current I_o through $R4$, $Q3$, $D1$, and R_L and determine the output voltage V_o by using I_o times the load resistance R_L. Alternately, a simple voltage divider calculation can be made as shown in Equation (3-3).

$$V_o = \left(\frac{R_L}{R_L + 130\ \Omega}\right)(5\ \text{V} - 0.8\ \text{V} - 0.2\ \text{V}) \qquad (3.3)$$

In most actual applications, when a gate is driving a load to ground, the load has a low enough impedance for some amount of base current to flow in $Q3$, but the impedence is not so low that the conditions of Equation (3-3) are met. Thus, in most grounded-load cases, Equation (3-2), which includes the I_b term, should be used to determine the output level. Grounded loads are not the normal TTL static load configuration, but when an output switches from a *low* to a *high* level, the transient load current flow is to ground (see Section 5-3).

Most TTL signals observed in the lab have *high* levels that are above 3.4 V and are often in the 4.0 to 4.7 V range. Yet Equation (3-2) indicates that the *high* limit is 3.4 V. Why the discrepancy? The inconsistency is due to the use of a load connected to ground in the model (Figure 3-14); normal TTL loads (inputs) tend to pull signals up rather than down. The actual load when TTL devices are driving other TTL devices is shown in Figure 3-15.

The load, in the normal case, the input of another TTL device, pulls V_o up to the supply level (5 V) less one diode forward-voltage drop (0.8

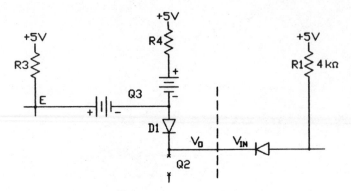

Figure 3-15   TTL output connected to TTL input.

V). The actual worst-case input-output current and voltage specifications for the various TTL families are shown in Chapter 2.

Equivalent circuit with *high* inputs.   The equivalent circuit for the case where both inputs are at a *high* level, where an input *high* level can range between 2 V and slightly above the plus supply rail (the input levels and limits are listed in Chapter 2), is shown in Figure 3-16. Both inputs must be at a *high* level to initiate the NAND action, i.e., a *low* output. With both inputs *high,* both input diodes $D2$ and $D3$ are reverse-biased and act as open circuits. Since the diodes act as open circuits, input levels have no influence on the voltage at node C. Node

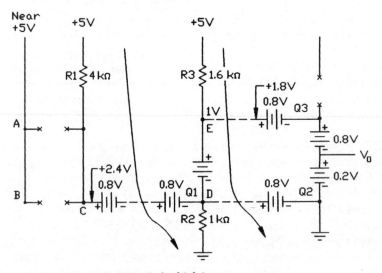

Figure 3-16   Equivalent circuit for *high* inputs.

C attempts to go toward +5 V but is clamped by the base-emitter junctions of $Q1$ and $Q2$. When node C attempts to go positive, it provides $Q1$ with a base current sourced from the +5-V supply through the 4-kΩ resistor. With base current, $Q1$ is turned ON; thus collector-emitter current flows in $Q1$. With $Q1$ ON and saturated and the output section disconnected, the voltage at node D is

$$V(\text{node } D) = \left(\frac{1 \text{ k}\Omega}{1 \text{ k}\Omega + 1.6 \text{ k}\Omega}\right)(5 \text{ V} - 0.2 \text{ V})$$

$$= 1.85 \text{ V}$$

and at node E is

$$V(\text{node } E) = V(\text{node } D) + 0.2 \text{ V}$$

$$= 2.05 \text{ V}$$

The next step is to connect the output section and establish that $Q2$ is ON and $Q3$ is OFF, the conditions required for the device to function as a NAND. For $Q2$ to be ON, it must have a source of base current, and it does since node D is at 1.85 V without $Q2$ connected. With $Q2$ connected to node D, node D is clamped to 0.8 V by the base-emitter junction of $Q2$. The clamping action of the base-emitter junction of $Q2$ changes the level at node E also; it becomes 0.8 V plus 0.2 V, or 1.0 V. It should be noted that the action of $Q2$ does not change the state of $Q1$. Transistor $Q1$ still has a source of base current, some of which may flow through the 1-kΩ resistor to ground while the remainder flows on to the base of $Q2$. Under these conditions transistor $Q2$ is ON since base current is provided through $Q1$ and the 1.6-kΩ resistor in the collector circuit of the phase splitter. With $Q2$ ON, the output V_o is pulled down to a level equal to the saturation voltage of $Q2$, which is typically near 0.2 V. Thus, the output is in a *low* state. A NAND gate with both inputs *high* is expected to have a *low* output.

In order to avoid a high-current condition in the output section, $Q2$ and $Q3$ must not be ON simultaneously. Thus, the next step is to determine that $Q3$ is OFF, which is best accomplished by assuming the conditions that are required for $Q3$ to be ON. That is, assume current is flowing through $D3$ so that the forward-voltage drop across $D3$ is 0.8 V and that current is flowing in the base-emitter junction of $Q3$ so that the base-emitter drop is 0.8 V. Then determine the voltage drops from the base of $Q3$ to ground. The result is 1.8 V, but node E, which is the same as the base of $Q3$, is at 1.0 V (see above analysis). Since 1.0 V is less than 1.8 V, and 1.8 V is necessary for $Q3$ to be ON,

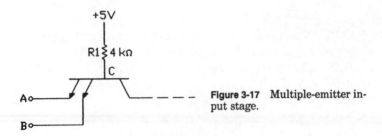

Figure 3-17 Multiple-emitter input stage.

$Q3$ must be OFF. Therefore, only one of the totem-pole output stage transistors is ON.

3-1-3 Multiple-emitter input structures

The operation of multiple-emitter inputs is not straightforward. Fortunately, most of the advanced Schottky families use diode input structures. However, some of the older families, in particular the original standard TTL family and the S or Schottky family, use multiple-emitter transistors (as shown in Figure 3-17) to implement input stages.[7] Under most conditions multiple-emitter input stages function in much the same manner as diode input stages. However, there are some exceptions; one of these is that multiple-emitter inputs are more sensitive to excessive *high* input-voltage levels.

The functional operation of a multiple-emitter input stage is unchanged by substituting diodes for the multiple-emitter transistor *pn* junctions, as shown in Figure 3-18. However, the electrical operation of a multiple-emitter input is different, and it is extremely important to never lose sight of the fact that the device is a multiple-emitter transistor and not several diodes tied together.

Two-input gate with multiple-emitter inputs. To examine the operation of a multiple-emitter two-input NAND gate, a multiple-emitter tran-

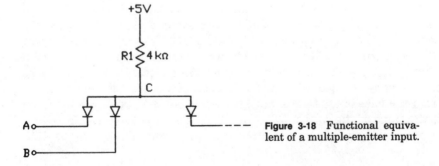

Figure 3-18 Functional equivalent of a multiple-emitter input.

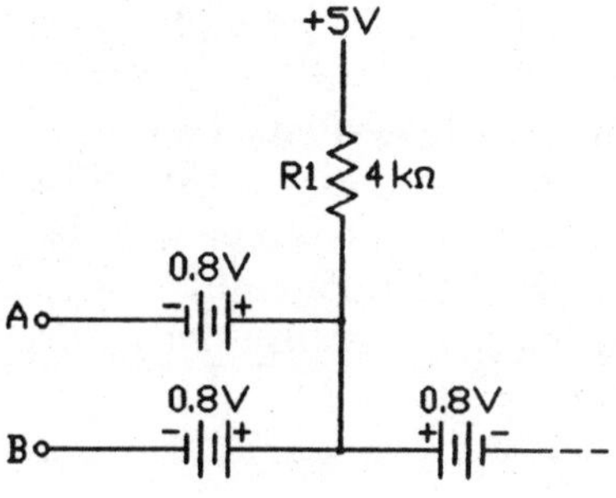

+5V

R1 ⪢ 4 kΩ

0.8V
Ao —=|||+

0.8V   0.8V
Bo —=|||+   +|||=— — —

Figure 3-19 *Low*-input equivalent circuit for a multiple-emitter input stage.

sistor is substituted for the diodes in the input section of the example gate of Figure 3-11; the remainder of the circuit is kept the same. Such a configuration is identical to the standard TTL NAND shown in Figure 3-1.

Multiple-emitter input with *low* inputs. Figure 3-19 shows the equivalent circuit for a multiple-emitter input stage with *low* inputs. The multiple-emitter *low*-input equivalent circuit (Figure 3-19) is the same as the previously discussed diode input circuit (Figure 3-12).

The circuit operation for a *low* input is as follows (see Figure 3-20): When either or both inputs are connected to a *low* level or to ground, the voltage drop across the base-emitter junction (or junctions) that is tied to ground is 0.8 V and has the polarity shown in Figure 3-20. With a *low* input, the current in the input section flows down through the 4-kΩ resistor and through the base-emitter junction of the grounded transistor to ground.

The phase-splitter and output-section circuitry function in the same manner as described earlier (for a refresher, refer back to Figure 3-12). For the lower transistor (*Q*2) in the output totem-pole stage to be

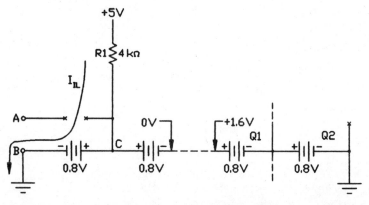

Figure 3-20   Equivalent circuit for multiple-emitter gate with a *low* input.

ON (see Figure 3-12), the three junctions between the grounded emitter of *Q2* and node *C* must be forward-biased. In actuality, none of the
junctions between node *C* and *Q2* are forward-biased since node *C* is
clamped to 0.8 V by the 0.8 V drop of the base-emitter junction of the
input transistor. Hence, *Q2* is OFF and the output is not in a *low*
state. Note that for the *low*-input case the multiple-emitter input
transistor does function as a group of diodes as shown in Figure 3-18.
However, that is not the case for *high* inputs.

REFERRING BACK TO THE TRANSISTOR MODEL SHOWN IN
FIGURE 3-7, WE SEE THAT IT DOES NOT APPEAR TO MAKE
MUCH DIFFERENCE WHICH TERMINAL IS CALLED THE COL-
LECTOR OR EMITTER, AND DEPENDING UPON A TRANSIS-
TOR'S CONSTRUCTION, IT MAY OR MAY NOT BE OF ANY CON-
SEQUENCE. ALL BIPOLAR TRANSISTORS HAVE SOME
INVERSE GAIN. IN MOST CASES THE GAIN IS OPTIMIZED FOR
THE NORMAL (OPERATING) CONFIGURATION, BUT IF THE
COLLECTOR AND EMITTER ARE INTERCHANGED, MOST
TRANSISTORS STILL HAVE A GAIN GREATER THAN 0 AND IN
MANY CASES MAY HAVE A GAIN GREATER THAN 1.

SOME OF THE EARLY, GROWN JUNCTION TRANSISTORS
HAD PHYSICALLY SYMMETRICAL BASE-EMITTER AND BASE-
COLLECTOR JUNCTIONS SUCH AS THE ONE SHOWN IN FIG-
URE 3-21.[3] SUCH TRANSISTORS HAD EQUAL GAIN IN EITHER
DIRECTION; THUS IT DID NOT MATTER WHICH TERMINAL
WAS CALLED THE COLLECTOR OR WHICH WAS CALLED THE
EMITTER; SUCH DEVICES FUNCTIONED IDENTICALLY IN EI-
THER CIRCUIT ARRANGEMENT. THE TRANSISTOR JUNC-
TIONS IN DIGITAL INTEGRATED CIRCUITS ARE NOT SYMMET-
RICAL, BUT THE TRANSISTORS DO HAVE SOME INVERSE
GAIN.

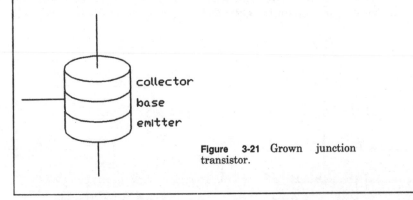

collector

base

emitter

Figure 3-21 Grown junction
transistor.

Multiple-emitter input with *high* inputs. For the *high*-input case, the
multiple-emitter input functions in a more conventional transistor

mode, except that the emitter and the collector functionally change places with respect to the schematic representation. Figure 3-22 shows the input-current path with one of the multiple-emitter inputs in a *high* state. If the action shown in Figure 3-22 is to occur, both inputs must be in a *high* state; if either one is *low*, the conditions of Figure 3-20 exist. Only one input is shown for simplicity. The current flow *I* shown in Figure 3-22 is reached when the input rises above a level where the clamping action of the diode junction on the input no longer holds node *C* below a level where no current flows into the base of the phase splitter *Q*1 and the output *Q*2 transistors. When that level occurs, the phase splitter and the bottom transistor in the totem-pole output stage turn ON and the output of the NAND gate goes to a *low* level. In this case, the voltage needed at the input is 1.6 V, as can be determined by an inspection of the circuitry in Figure 3-23. Above 1.6 V, (1) the input transistor changes to the inverted mode, (2) the terminal drawn as a collector in Figure 3-17 becomes the emitter, and (3) the terminals drawn as emitters start functioning as collectors as shown in Figure 3-22.

When the inputs are *high,* current is supplied to the phase splitter by the input-stage resistor *R*1 and the input signals. The resistor *R*1 supplies most of the current, as is also the case when the input circuit consists of true diodes. If the inverse current gain of the input multiple-emitter transistor is low, as it should be, then most of the current to the phase splitter is supplied by *R*1 through the base connection and the signal sources (inputs) are not required to supply much current. However, if input levels are higher than normal, significant input current may flow depending upon the inverted mode

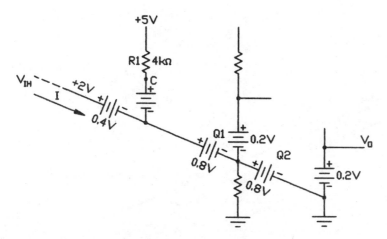

Figure 3-22   Multiple-emitter input with a *high* input.

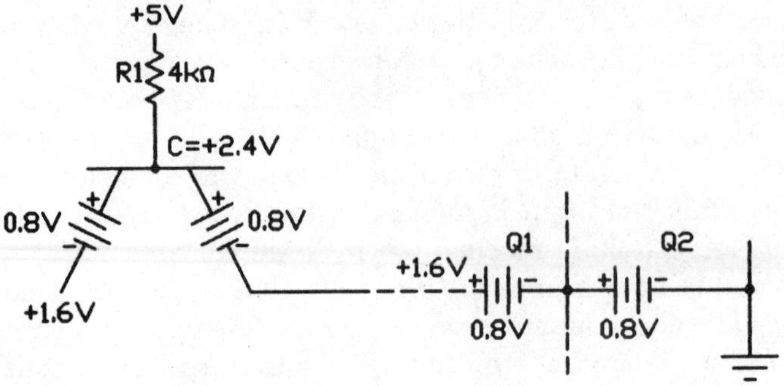

Figure 3-23   Threshold for multiple-emitter input stage.

gain of the multiple-emitter input transistor. Figure 3-24 is the equivalent circuit of a multiple-emitter input connected to +5 V. The equivalent circuit (Figure 3-24) shows that there is no current-limiting resistance in the collector-emitter path and that the internal device voltage drops do not equal the input source (+5 V). Excessive current flow is to be expected when such a condition exists.

For example, transistors should never be connected, as shown in Figure 3-24 or 3-25, with no collector or emitter current-limiting when there is a sufficient source of base current to drive the collector to a high current level. The life expectancy of a transistor connected as shown in Figure 3-25 is not very long. Yet, that is the connection when a multiple-emitter input is tied directly to the +5-V supply.

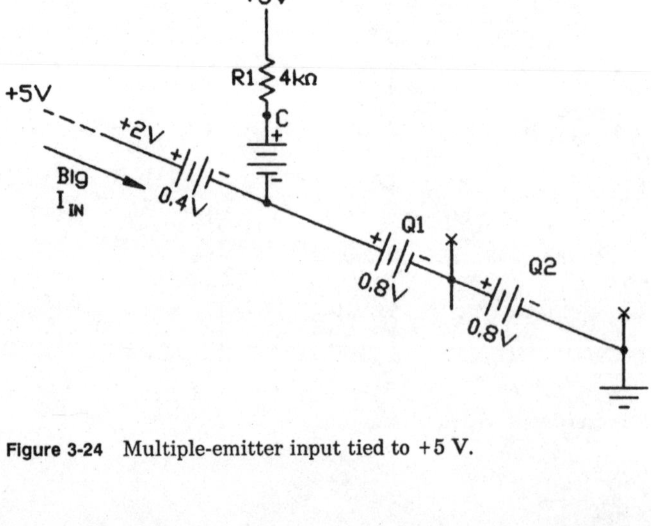

Figure 3-24   Multiple-emitter input tied to +5 V.

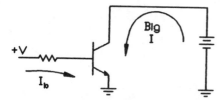

Figure 3-25 Transistor with no collector-emitter current-limiting.

In addition to the inverted input transistor input current path shown in Figure 3-24, multiple-emitter inputs have inverted transistor modes that are a function of input signal polarity. These additional inverted modes further complicate matters when devices with multiple-emitter inputs are used. Figure 3-26 shows a multiple-emitter input stage with one input *high* and one input *low*. When the signal polarities to a multiple-emitter input result in the levels shown in Figure 3-26, the emitter connected to the *high* input becomes a collector, but the emitter connected to the *low* input continues to function as an emitter. The resulting transistor is shown in Figure 3-27.

The transistors formed as a result of *high* and *low* inputs are usually not very good transistors. Manufacturers try to design the multiple-emitter input transistors so that emitter-to-emitter transistors have little gain. However, in all cases multiple-emitter input stages must be treated as transistors, not as diodes. Multiple-emitter inputs will have excessive current flow if the input levels exceed normal TTL levels (2.5 to 3.5 V) and there is no current-limiting. In normal TTL device-to-device interconnection situations, device inputs are driven by similar devices and the output structures of all TTL devices contain current-limiting resistors in the path between the output and the supply voltage V_{cc}. Thus, under normal conditions TTL *high*-output signals are current-limited and *high* input-voltage levels are less than +5 V because of the base-emitter drop of the output pull-up

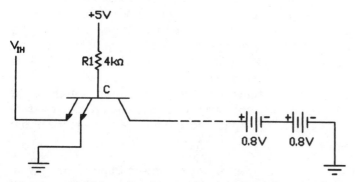

Figure 3-26 Multiple-emitter input with one input *high* and one *low*.

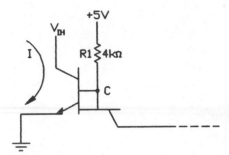

Figure 3-27 Transistor formed when one input is *high* and one is *low*.

transistor and the drop across the output diode. Thus, in normal operation, multiple-emitter input transistors are not destroyed when inputs go *high*. However, when multiple-emitter inputs are driven by other than TTL sources, care must be taken to provide adequate input current-limiting.

Fortunately, most of the advanced Schottky TTL families no longer use multiple-emitter inputs. In most cases the newer families use real diodes for the input structure, and real diodes are much easier to deal with. Since most manufacturers no longer supply equivalent schematics for most devices, how does one tell whether a device has multiple-emitter inputs? A clue is provided in the manufacturer's absolute maximum input-voltage specifications. If the input circuits are made with real diodes, the absolute maximum input *high* specification is usually 7.0 V. If multiple-emitter inputs are used, the absolute maximum input specification is 5.5 V. Since the absolute maximum rating for multiple-emitter inputs is 1.5 V less than that of diode inputs, directly connecting (with no current-limiting) unused multiple-emitter inputs to $+V_{cc}$ reduces by 1.5 V the tolerance of the system to transients or other over-voltage conditions on the V_{cc} supply. A 1.5-V reduction in the ability of low voltage devices to withstand unusual conditions is a severe penalty. Thus, unused inputs of multiple-emitter devices that must be connected to a *high* level for proper device operation must not be tied directly to +5 V (V_{cc}). Some type of current limited *high* logic level (such as a resistor to +5 V) must be used. If current-limiting is not provided, small positive-going power supply transients that exceed input-voltage ratings but do not exceed supply ratings may destroy inputs without damaging the core of devices. However, most parts are of little use if the inputs are destroyed. If we expect to build reliable systems, we must take full advantage of the margin built into the devices that we use and not compromise the ruggedness of the systems by failing to take advantage of the margin that exists.

Pull-up resistors. For most unused input pull-up applications for either TTL or CMOS devices, 1-kΩ resistors are suggested. Higher values increase the chance for noise pickup (due to higher source imped-ance) and the chance for problems due to board leakage currents or defective devices with excessive input currents. The drop across pull-up resistors must not be large enough to cause the voltage on the load side of pull-ups to be less than the minimum *high*-input level plus some allowance for noise margin. That is,

$$V_{cc\ min} - (I_{in\ total})(R_{pull-up}) \geq V_{in\ high\ min}$$

Since TTL *high*-input currents are very low, 20 μA, and even lower for CMOS, 1 μA, a large number of inputs can be pulled up with one pull-up resistor without violating minimum *high*-input level require-ments. However, it is generally not wise, nor practical, to tie too many points together. Troubleshooting a shorted line tied to a great number of points can be very difficult, and the longer the line, the greater the chance for coupled noise. Thus, for practical considerations, the max-imum number of inputs pulled up by one resistor should be limited to 10.

Additional sources of current-limited pull-up or pull-down voltages. Other connections that can be used to tie off unused inputs include:

1. Tie unused inputs requiring *high* inputs to *high* outputs of invert-ers that have their inputs tied to ground.
2. When current-limited *low* levels are required, tie unused inputs re-quiring *low* inputs to outputs of inverters or noninverting buffers that have their inputs tied to the appropriate levels.
3. Tie unused inputs to used inputs that are functionally the same (on the same device).

3-1-4 Paralleling inputs

When unused inputs are tied to used inputs that are functionally the same (on the same device), the logical operation of the device is not affected. However, paralleling inputs can degrade dynamic perfor-mance and may affect dc loading. The capacitance load of the input signal is increased, which can adversely affect high-frequency perfor-mance and reduce the noise margin of devices with diode input struc-tures. Paralleling inputs parallels the capacitance of the input diodes and increases the capacitances between the input node and the inter-nal structure of the device. The increased capacitance increases the possibility that noise on the input signal will be injected into the in-ternal circuitry of the device and upset it. Thus, paralleling inputs is not recommended for LS devices[9] (and should be avoided for most de-vices with diode inputs).

Parallel input loading.   Paralleling inputs of a given gate results in different unit loading for the driving device for the *low-* and the *high-*input state. For the *low-*input state, paralleled inputs present only one unit load to the driving device regardless of the number of inputs paralleled (the case being described pertains to standard gates and inputs common to a given gate only). Figure 3-28 shows the equivalent circuit of two paralleled inputs. The only change to the circuit, which is a result of paralleling the inputs, is a paralleled equal-value voltage source. Thus, paralleling *low* inputs has no effect on the current that flows to the driver. Paralleling one or *N* inputs (of a given device) results in the same *low-*input current since paralleling equal-value voltage sources (the equivalent circuit for a forward-biased diode) causes no change in the current flow in the circuit.

However, when inputs are paralleled and the input is *high,* the load is equal to the number of inputs paralleled. Each input appears as an individual current source (sink) as shown in Figure 3-29.

3-2   CMOS Circuits

CMOS logic devices are built with enhancement-mode metal oxide semiconductor field-effect transistors (MOSFETs) arranged in complementary *n*-channel and *p*-channel pairs.[10] The complementary pairs use *p*-channel MOSFETs for pull-ups and *n*-channel MOSFETs for pull-downs. Logic devices implemented with complementary pairs have no dc paths between V_{cc} and ground; one switch in each complementary pair is ON and the other OFF at all times. Hence, CMOS devices dissipate little dc power. A typical complementary inverter stage is shown in Figure 3-30.

Most AC and HC logic devices are fabricated with complementary

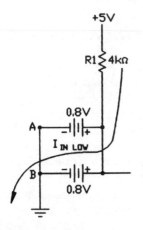

+5V

R1 ⪤ 4kΩ

0.8V

A

$I_{\text{IN LOW}}$

B

0.8V

Figure 3-28 Load when inputs are paralleled and the input is *low.*

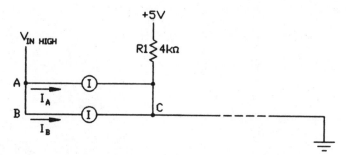

Figure 3-29 Multiple-emitter gate with inputs paralleled and a *high* input.

silicon oxide semiconductor material rather than complementary metal oxide semiconductor material, but they are still referred to as CMOS devices.

3-2-1   Enhancement-mode field-effect transistors

The symbols and terminal nomenclature for enhancement-mode field-effect transistors (FETs) are shown in Figure 3-31.[3]

Enhancement-mode n-channel FETs operate somewhat analogously to *npn* bipolar transistors, and p-channel FETs operate somewhat analogously to *pnp* bipolar transistors (see Figures 3-32 and 3-33).

3-2-2   FET equivalent circuits

The equivalent circuit for an ON MOSFET is a resistor and the equivalent circuit for an OFF MOSFET an open circuit (see Figure 3-34).

Typical ON resistance of advanced CMOS devices is near 10 Ω and

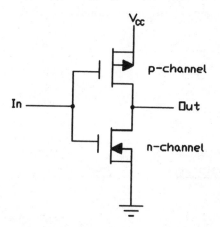

Figure 3-30 Typical CMOS inverter stage.

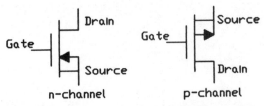

Figure 3-31 Enhancement-mode field-effect transistors.

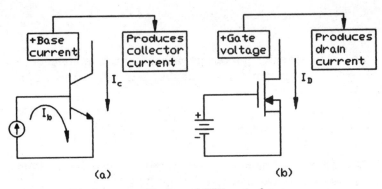

Figure 3-32 Bipolar *npn* and *n*-channel FET equivalency.

that of HC logic components is near 100 Ω. The OFF impedance of most FETs is extremely high; OFF FETs have some leakage current, but in most digital applications OFF leakage currents are of such low magnitude that OFF FETs can be considered open circuits.

Figure 3-35 shows a simple CMOS inverter and its equivalent circuit,[7] which consists of two resistors and two ideal switches in series between V_{cc} and ground. In the equivalent circuit, the switching action of the FETs is represented by the ideal *p*-switch and *n*-switch.

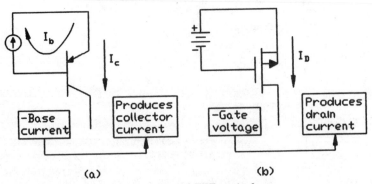

Figure 3-33 Bipolar *pnp* and *p*-channel FET equivalency.

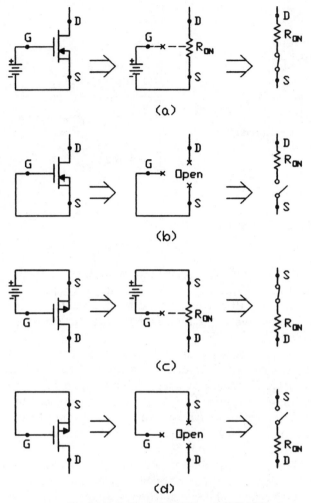

Figure 3-34 Equivalent circuits for ON and OFF FETs.

The two resistors represent the ON resistance of the two FETs. The inverter functions as follows: When the input level is near V_{cc}, the p-channel pull-up FET is OFF and the n-channel pull-down FET is ON. When the input level is near ground, the n-channel pull-down FET is OFF and the p-channel pull-up FET is ON. When the n-channel pull-down FET is ON, the output level is near ground. When the p-channel pull-up FET is ON, the output level is near V_{cc}.

3-2-3 CMOS logic circuits

Logic functions, such as NORs and NANDs, are implemented using multiple complementary structures.[3,7] For example, two-input CMOS

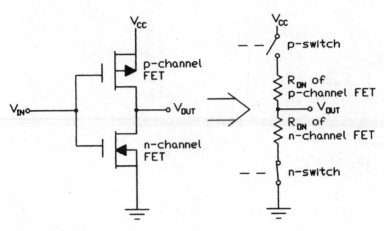

Figure 3-35   Inverter and equivalent circuit.

NOR gates are implemented with two series pull-up and two parallel pull-down FETs as shown in Figure 3-36. The opposite arrangement, parallel pull-ups and series pull-downs, is used to implement CMOS NAND gates. For example, two-input NAND gates are implemented as shown in Figure 3-37.

To implement logic functions with multiple inputs, the number of complementary pairs of series and parallel FETs can be increased as needed (up to some practical limit). However, as the number of series and parallel devices increases, output *high* and *low* impedances become more unbalanced and devices with large numbers of inputs have very unsymmetrical responses. All advanced CMOS devices have output buffers to balance output drive and response. Output buffers isolate internal logic function structures from line and load capacitances, which tend to be much larger than internal device node capacitances. Some of the early 4000 series CMOS logic functions are unbuffered

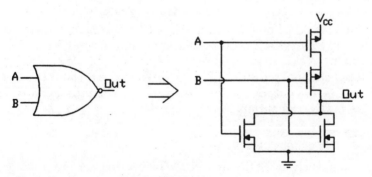

Figure 3-36   Two-input CMOS NOR gate.

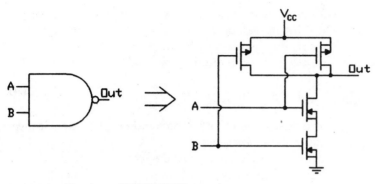

Figure 3-37 Two-input CMOS NAND gate.

and have very unsymmetrical responses. The variable output impedance of an unbuffered two-input NOR gate (Figure 3-36) is shown in Figure 3-38.

3-2-4 Dynamic power dissipation

Calculating the power dissipation of a CMOS system is extremely difficult. The majority of the power dissipated in a CMOS system is a function of dynamic conditions in contrast to TTL systems where most power dissipation is a function of intrinsic device dissipation. System dc power dissipation is easy to calculate from device dc specifications. System dynamic power dissipation, on the other hand, is difficult to calculate since dynamic power is dependent on the system application rather than on the inherent device characteristic. Dynamic power

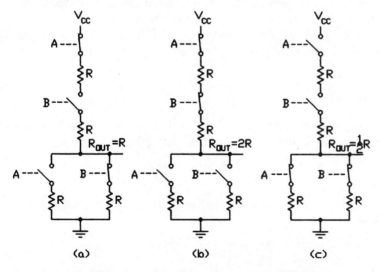

Figure 3-38 Variable output impedance of a two-input NOR gate.

dissipation is a function of node frequency, capacitance, and signal voltage swing. Thus, the operating conditions of each system signal, or node, is needed to calculate system dynamic power.

The equation for calculating dynamic power is[11]

$$P_d = (C_L + C_{pd})(\Delta V_s)^2 F \qquad (3.4)$$

where C_L = line and device load (both input and output) capacitance
C_{pd} = internal device capacitance
ΔV_s = signal swing ($\approx V_{cc}$ for CMOS devices)
F = node toggle frequency

Signal voltage swing is easily established, and C_{pd} is provided on most CMOS data sheets, but line and load capacitance and the frequency of each node are usually difficult to establish. Yet, if a reasonably accurate estimate of CMOS system power is needed, a reasonably accurate estimate of the toggle rate and capacitance of each node is needed to calculate system power dissipation. In most situations, the best that can be done is to neglect all low-frequency control signals and concentrate on data paths where some assumptions as to average toggle rate and load capacitance can be made.

It is generally assumed that CMOS systems dissipate less power than equivalent TTL systems. Perhaps that is true at low frequencies since CMOS devices dissipate little dc power, but as system operating speeds increase, the power advantage of CMOS systems decreases. Since CMOS devices have greater output-voltage swings than TTL devices, dynamic power dissipation is greater in CMOS systems than in TTL systems [see Equation (3-4)]. Typical advanced CMOS logic device output-voltage swings are 5 V, whereas typical TTL output-voltage swings are 3 V. Thus, typically, CMOS dynamic power is greater than TTL dynamic power by a factor of

$$\frac{5^2}{3^2} = \frac{25}{9} \approx 2.8$$

For example, the dynamic load power dissipation for a CMOS-level signal driving a 50-pF load at 10 MHz is [using Equation (3-4)]

$$P_d = (50 \text{ pF})(5 \text{ V})(10 \text{ MHz})$$
$$= 12.5 \text{ mW}$$

and for the same load and signal switching frequency, TTL signal dynamic power is

$$P_d = (50 \text{ pF})(3 \text{ V})(10 \text{ MHz})$$
$$= 4.5 \text{ mW}$$

Thus, if all signals in a system toggle at a high rate, the use of CMOS technology may provide little or no power advantage over TTL. Yet, most high-speed CMOS systems dissipate less power than equivalent TTL systems because in most systems many of the control signals toggle at relatively low rates; usually only a few signals toggle at high rates and have high dynamic-power dissipation. However, if all signals toggle at very high rates, for example, 40 MHz or greater, the dissipation of a CMOS systems may exceed the dissipation of an equivalent TTL systems.

3-2-5 Input protection

Isolated gate FETs, such as those used in CMOS IC logic devices, are highly susceptible to damage from electrostatic buildup unless some form of low-impedance bypass circuitry is added to exposed external device terminals.[12] Isolated gate FETs without bypass circuitry have extremely large gate-to-drain and gate-to-source impedances, typically on the order of 10^{12} Ω. Such high-impedances levels are conducive to static buildup and must be avoided between external points (i.e., package pins) if electrostatic buildup is to be avoided.

To prevent damaging levels of electrostatic charge, CMOS ICs have low-impedance discharge paths made up of various combinations of diodes (see Figure 3-39) on all input-output connections. The diodes are arranged to clamp either positive or negative electrostatic voltage excursions to levels below FET gate oxide breakdown limits. In addition to diodes, most input protection networks include some resistance in order to limit the current in the clamp diodes, so as to protect the diodes, but most output clamps do not have current-limiting resistors (see Figure 3-39). Input- or output-protection networks have little effect on signals when signal levels are between V_{cc} and ground, but when signal levels exceed V_{cc} or ground, some clamping action may occur depending on the speed of response of the diodes. Input- or output-protection diodes should not be relied on to provide effective

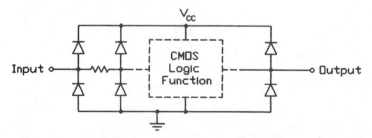

Figure 3-39 Typical CMOS electrostatic protection circuits.

clamping of high-frequency signals, although in some cases, input-protection networks do provide effective dynamic clamping. In most instances, however, input-protection network impedance is too high and diode response too slow for effective dynamic signal clamping. Different manufacturers use different protection networks for the same generic part, so care must be exercised if the dynamic response of protection networks is important. Each manufacturer's parts must be evaluated for the particular application at hand. In all applications, large static and dynamic currents must be prevented from flowing in input and output networks to prevent possible latch-up of the parasitic SCRs that are inherent in CMOS integrated circuits (see Section 3-2-6).

Board or unit CMOS interface protection.   Isolated board or system interface signals that connect directly to CMOS inputs or outputs must be avoided. All external board or unit signals that connect to CMOS inputs should have shunt low-impedance paths to V_{cc} or ground and series current-limiting resistors (see Figure 3-40).[10] Shunt low-impedance paths help prevent electrostatic buildup when boards or units are isolated (e.g., when a board is out of the chassis). When a unit is in place and powered, shunt resistors prevent open inputs from floating. All inputs and outputs should have series current-limiting resistors to prevent excessive input or output currents. Series current-limiting resistors provide a means of controlling static and transient currents injected into protection networks. Signals that overshoot or undershoot V_{cc} or ground can induce latch-up if excessive current is allowed to flow in protection networks. Series resistors are also necessary to limit current in protection networks in applications where signal sources may be powered when receiving devices are not. In all CMOS interface applications, input currents under abnormal or worst-case transient conditions must be kept below actual, specified device limits. Absolute maximum dc input-current limits for most CMOS logic devices are in the 20- to 30-mA range. Transient limits

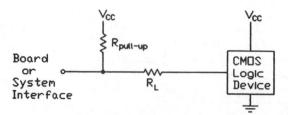

Figure 3-40   Board or unit CMOS device input protection.

for most advanced CMOS devices are typically near 100 mA, but caution must be exercised because some CMOS devices may latch-up
when the injected current is as low as 10 mA.

In applications where high-speed signals must cross external board
or unit boundaries, series current-limiting resistors may cause excessive RC delays. However, if it is not possible to use series current-
limiting resistors and the possibility of excessive static or transient input currents exists, other steps must be taken to control the situation.

3-2-6 Latch-up

All CMOS devices are intrinsically susceptible to latch-up,[13,14] which
occurs when internal parasitic silicon-controlled rectifiers (SCRs),
structures that are inherent in CMOS integrated circuits, are triggered ON. When triggered ON, parasitic SCRs cause low-impedance
paths between V_{cc} and ground that remain ON until V_{cc} is removed or
the part is destroyed. Latch-up usually disrupts the functional operation of a part and in many cases will cause permanent damage even
though some parts may return to normal operation after power is cy-
cled OFF and then back ON. Latch-up may be initiated by voltage
overshoots or undershoots that cause substrate currents that exceed
device ratings at one or a combination of device inputs, outputs, or
supply terminals.

All CMOS integrated circuits have parasitic four-layer *pnpn* structures, shown in Figure 3-41, that can be triggered into a regenerative
switching mode if sufficient current is injected into the appropriate
points. Four-layer structures associated with input-output circuits are
most exposed to transient currents and are most likely to latch-up.
Manufacturers use a number of techniques to minimize the chance for
latch-up: increased spacing between parasitic devices to reduce gain,
guard rings around diffusion areas, and low-impedance substrates to
prevent injected currents from developing sufficient potential to trigger parasitic structures, to name a few. However, all CMOS integrated circuits have parasitic SCR structures that can be triggered if

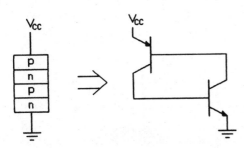

Figure 3-41 Four-layer SCR
structures found in CMOS integrated circuits.

sufficient current is injected. Most advanced CMOS logic devices have latch-up immunity for injected current levels of 100 to 200 mA, depending upon device and device manufacturer. Some of the older devices may latch-up with injected current levels as low as 10 mA. High temperature increases the susceptibility of devices to latch-up. Thus, parts should be kept as cool as possible, and extra precautions must be taken to limit substrate input-output currents at high temperature. Systems that operate without latch-up at room temperature may latch-up at high temperature because of injected currents from normal signal overshoots. Inputs must not be allowed to float since floating inputs, or inputs with slow transitions, increase device operating temperature and increase the chance for latch-up. Unused CMOS inputs should not be connected directly to V_{cc} or ground; some current-limiting is needed.

Latch-up can be prevented by limiting both static and transient input and output currents. Where signals cross board or system unit boundaries and are most exposed to transient conditions, current-limiting should be provided to ensure that device current and voltage ratings are not exceeded. Where CMOS sources and loads are not powered by the same source, interconnecting signal lines must be current-limited or clamped to prevent injected substrate currents in unpowered parts. Unpowered parts with substrate currents caused by powered input signals may latch-up when local V_{cc} is applied.

Series resistors provide the simplest means of current-limiting CMOS interfaces. Current-limiting resistors should be as large as possible, but resistor size must be balanced against excessive speed degradation. Under most board-to-board and other internal system conditions, special current-limiting is not required when advanced CMOS devices are used. However, if long external lines with a high probability of being exposed to abnormal conditions must interface with CMOS devices, or if some of the older CMOS logic families or special custom devices are being used, great care must be taken to limit interface currents.

References

1. Grove, A. S.: *Physics and Technology of Semiconductor Devices*, Wiley, New York, 1967.
2. van der Ziel, Aldert: *Solid State Physical Electronics*, Prentice-Hall, Englewood Cliffs, N. J., 1968.
3. Millman, Jacob: *Microelectronics, Digital and Analog Circuits and Systems*, McGraw-Hill, New York, 1979.
4. Millman, Jacob, and Herbert Taub: *Pulse, Digital, and Switching Waveforms*, McGraw-Hill, New York, 1965.
5. Lin, H. C.: *Integrated Electronics*, Holden-Day, San Francisco, Calif., 1967.
6. Holt, Charles A.: *Electronic Circuits Digital and Analog*, Wiley, New York, 1978.

7. Mano, M. Morris: *Digital Design,* Prentice-Hall, Englewood Cliffs, N. J., 1984.
8. *FAST Advanced Schottky TTL Databook,* National Semiconductor Corp., Santa Clara, Calif., 1988.
9. *Bipolar Microprocessor Logic and Interface Data Book,* Advanced Micro Devices Inc., Sunnyvale, Calif., 1981.
10. Walsh, M. J.: *Choosing and Using CMOS,* McGraw-Hill, New York, 1985.
11. *System Power Models for HCMOS Devices,* SCLA003, Texas Instruments Inc., Dallas, Tex., 1985.
12. McAteer, Owen J.: *ESD Control: Principle and Practices,* McGraw-Hill, New York, in press.
13. Wakeman, Larry: "Closing in on CMOS Latch-Up," *Integrated Circuits Magazine,* April 1985, pp. 38–44.
14. Troutman, Ronald R.: *Latchup in CMOS Technology,* Kluwer Academic Publishers, Hingham, Mass., 1986.

4

Noise Margin

Noise margin is the difference between the worst-case output level of a driving device and the worst-case input level at which the receiving device no longer recognizes the input as the intended *high* or *low* logic level. For example, Figure 4-1 illustrates Schottky TTL input-output limits and noise margins. Notice that *high* and *low* noise margin are different. Most TTL devices and most CMOS devices with TTL input levels have more *high* than *low* noise margin. Most CMOS devices with "true" CMOS inputs have equal *high* and *low* noise margins.

Noise margin is of great concern when advanced Schottky or advanced CMOS devices are used; their fast edges generate a tremendous amount of noise. Yet, their noise margins are approximately the same as those of the older, slower devices (in the respective technology) that generate much less noise.* Advanced CMOS devices have more noise margin than TTL devices, but they need more noise margin since their wider, and in many cases faster, voltage swings generate more noise.

4-1 Noise Margins of TTL Logic Families

Worst-case noise margins over the military temperature range for the various TTL families are listed in Table 4-1. The worst-case *high* and the worst-case *low* noise margins are different for all the families except the original TTL family, where both are 400 mV. For all the Schottky families, *low* noise margin is 300 mV and *high* is 500 mV. A 200-mV difference between *high* and *low* noise margin may not seem significant, but advantage should be taken of higher noise margin

---

*Advanced Schottky devices have higher threshold levels than some of the older TTL devices, but their worst-case noise margins are the same as those of the older Schottky devices (see Reference 1).

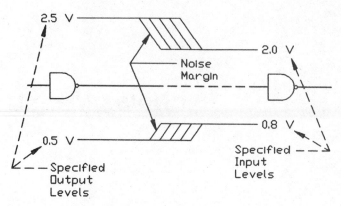

Figure 4-1 Schottky TTL noise margins.

where it exists. In many applications, 200 mV of additional noise margin will significantly enhance system reliability. Chapters 8 and 9 describe several means of taking advantage of the higher noise margin of Schottky *high*-level signals.

The noise margins shown in Table 4-1 are derived from worst-case specified input- and output-voltage levels for the various families (see Table 2-3 and References 1–5). Since the input-output voltage levels are the same for most of the Schottky TTL families, the noise margins are also the same. The exception is the LS family; LS *low* input-output levels are 0.7 and 0.4 V versus 0.8 and 0.5 V for the other Schottky families. However, since the difference is the same, LS *low* noise margin is the same as that of the other Schottky families when LS parts drive other LS parts.

4-1-1 TTL noise margin cross-reference tables

Tables 4-2 and 4-3 list noise margins for various combinations of interconnections among the various TTL logic families. Table 4-2 lists *low* noise margin and Table 4-3 lists *high* noise margin. Notice that minimum *low* noise margin (200 mV) occurs when the other Schottky

TABLE 4-1   Noise Margins of TTL Families

Family	*Low*-level noise margin, mV	*High*-level noise margin, mV
(TTL)	400	400
LS	300	500
ALS	300	500
F	300	500
S	300	500
AS	300	500

TABLE 4-2   TTL *Low*-Level Noise Margin Cross-Reference (in millivolts)

From	(TTL)	LS	ALS	F	S	AS
			To			
(TTL)	400	400	400	400	400	400
LS	400	300	400	400	400	400
ALS	300	200	300	300	300	300
F	300	200	300	300	300	300
S	300	200	300	300	300	300
AS	300	200	300	300	300	300

TABLE 4-3   TTL *High*-Level Noise Margin Cross-Reference (in millivolts)

From	(TTL)	LS	ALS	F	S	AS
			To			
(TTL)	400	400	400	400	400	400
LS	500	500	500	500	500	500
ALS	500	500	500	500	500	500
F	500	500	500	500	500	500
S	500	500	500	500	500	500
AS	500	500	500	500	500	500

families are used to drive LS devices. Also, notice that for all possible interconnection combinations, except for signals originating from standard TTL devices, worst-case *high* level noise margin is 500 mV.

4-1-2   TTL noise margin change with temperature

The threshold of TTL devices (i.e., the actual level where they switch states) is determined by the forward-voltage drops across a combinations of *pn* or metal-silicon junctions referenced to ground. For example, the junctions that determine the threshold for a standard (original) TTL NAND gate are shown in Figure 4-2. The forward-voltage drop across each of the junctions increases with lower operating temperatures and decreases with higher operating temperatures.[*] It follows that the threshold shifts in a similar manner. Thus, the actual noise margins, which are related to the actual input threshold level, shift with temperature.[6] *Low* noise margin increases with low temperature and *high* noise margin decreases with low temperature. The opposite happens under high operating temperature conditions: *low* noise margin decreases and *high* increases. Data books only show

---

[*]The temperature coefficient for the forward-voltage drop across a *pn* junction or a metal-silicon (Schottky) diode junction is approximately −2 mV/°C (see Reference 7).

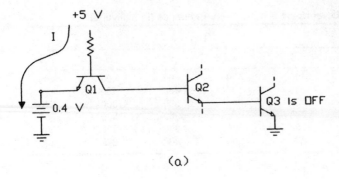

(a)

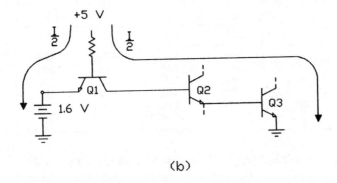

(b)

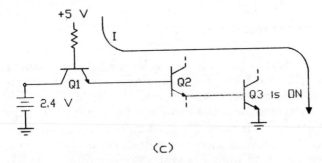

(c)

Figure 4-2   TTL NAND threshold-setting junctions.

worst-case *high* and *low* input limits, but designers should be aware of how the actual input threshold changes with temperature. In some cases, signal polarity can be arranged to take advantage of the actual (difference) noise margin.

4-1-3   Circuit models for determining TTL noise margin

In this section the circuit modeling and analysis techniques of Chapter 3 are used to determine the *high* and the *low* noise margins of two common TTL input structures for the two extremes of the military operating temperature range.

The first step is to determine the drop across a *pn* junction at the two temperature extremes, $+125$ and $-55°C$. At $-55°C$, the change in the forward-voltage drop ΔV_f of a *pn* junctions is

$$\Delta V_f = \text{(temperature coefficient)}(\Delta T)$$

$$= (-2 \text{ mV/°C})[-55°C - (+25°C)] \tag{4-1}$$

$$= (-2 \text{ mV/°C})(-80°C)$$

$$= +160 \text{ mV}$$

Using 800 mV (0.8 V) for the nominal forward drop at $+25°C$, the forward drop at $-55°C$ is

$$V_f = 800 \text{ mV} + 160 \text{ mV}$$

$$= 960 \text{ mV}$$

At $+125°C$ the change in the forward drop of a *pn* junction is

$$\Delta V_f = \text{(temperature coefficient)}(\Delta T)$$

$$= (-2 \text{ mV/°C})(125°C - 25°C) \tag{4-1}$$

$$= (-2 \text{ mV/°C})(100°C)$$

$$= -200 \text{ mV}$$

The forward drop at 125°C is (assuming a nominal 800-mV diode drop)

$$V_f = 800 \text{ mV} - 200 \text{ mV}$$

$$= 600 \text{ mV}$$

Circuit model for determining *high* noise margin.   *High* noise margin is worst at $-55°C$ since junction drops are maximum at low temperature. Increased junction drops shift the threshold in the positive direction and decrease the difference between the specified minimum *high* input level and the actual threshold. The circuit model for determining *high* noise margin for a standard TTL NAND (refer to Figure 3-1) is shown in Figure 4-3. The voltage at node *C*, which is a function of input level, determines whether the output is *high* or *low*. With 960-mV (0.96-V) *pn*-junction drops, node *C* must be 2.88 V (3 *pn* junctions at 0.96 V each) or above for current to flow in the direction that turns the lower transistor in the output stage ON. With the input at 2.4 V, which is the specified minimum *high* signal level for standard TTL devices (refer to Chapter 2 for input-output level specifications), the voltage at node *C* in Figure 4-3 is 3.36 V (2.4 V + 0.96 V). Thus, the input

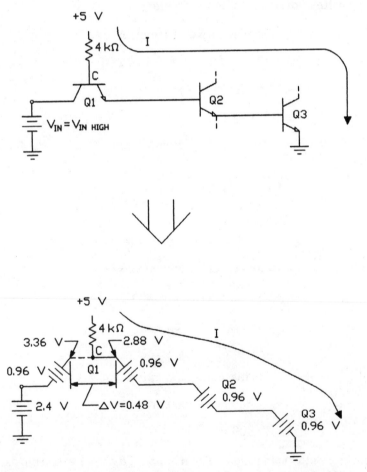

Figure 4-3 Circuit model for determining *high* input noise margin at −55°C.

can drop 0.48 V (3.36 V − 2.88 V) before current flow to the lower output transistor ($Q3$) ceases. The allowable drop in input level, 0.48 V, is the safety margin, or the noise margin. Specified worst-case *high* noise margin for standard TTL devices is 0.4 V (400 mV). The calculated value exceeds the worst-case value, as should be expected. The simple model should not be expected to produce worst-case results.

Circuit model for determining *low* noise margin. *Low* noise margin is worst at high temperature since junction drops decrease as temperature increases. Decreased junction drops shift the threshold in the negative direction and decrease the difference between the specified minimum *low* input level and the actual threshold. Figure 4-4 shows the circuit model for determining *low* noise margin for a standard TTL NAND. At +125°C, the forward-voltage drop for the various *pn* junctions is 600 mV (0.6 V). With a worst-case maximum *low* input of 0.4 V (see Chapter 2), the voltage at node C is 1.0 V. At +125°C, node C must reach 1.8 V for input stage current I to change directions and flow toward the output stage (which must occur for the gate to switch states—see Figure 4-4). Thus, the safety margin, or *low*-input noise margin, is 0.8 V (1.8 V − 1.0 V) at +125°C. Specified worst-case *low* noise margin for the original TTL family is 0.4 V (see Table 4-1).

A similar approach can be used to determine the approximate noise margin for the other TTL families. Noise margins determined with simple models are not worst case since the models do not represent worst-case conditions. However, noise margins determined with the aid of the simple models are representative of typical conditions. Manufacturer's worst-case specifications must have some allowance for variations in process and environmental conditions; thus the specified worst-case values for noise margin are expected to be smaller than the derived values.

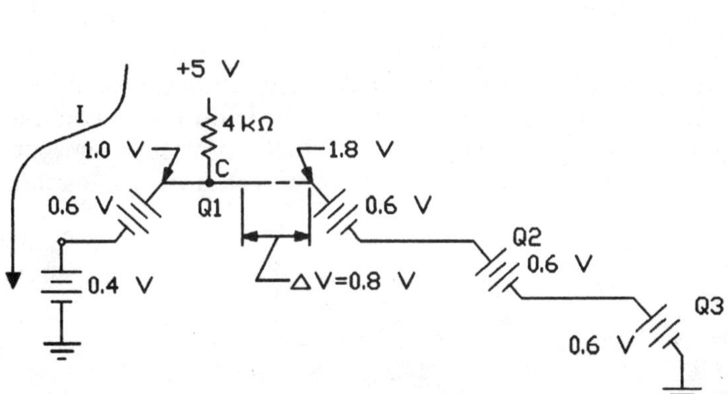

Figure 4-4 *Low* input noise margin at +125°C.

4-1-4 Summary of TTL noise margin considerations

An understanding of TTL noise margin levels and how noise margins change with temperature is needed to optimize TTL digital systems so that maximum advantage is taken of the existing noise margin. For example, Schottky TTL devices have more *high*-level noise margin than *low*-level noise margin. Designers should take advantage of the higher *high*-level TTL noise margin and arrange the polarity of critical signals that have low duty cycles so that the signals are in the *high* state most of the time. Thus, critical control signals, such as resets, should be *high* when inactive. Only devices that clock on *low-to-high* transitions should be used so that clock signals are *high* during the noisy time immediately following clock-switching edges.

If a system operates at a given temperature most of the time, critical signals should be arranged so that they are in the actual maximum noise-immunity state at critical times. For example, actual *high*-level Schottky TTL noise margin is greatest at elevated operating temperatures, and most systems operate at elevated temperatures most of the time.

Techniques for optimizing system noise immunity are covered in Part 2.

4-2 Noise Margins of CMOS Logic Families

In general, CMOS devices have greater noise margins than TTL devices.[7] Advanced CMOS devices with TTL input thresholds have improved *high*-level noise margin but have approximately the same *low* noise margin as advanced Schottky devices. Advanced CMOS devices with "true" CMOS input levels do have greater noise margins than TTL devices, but advanced CMOS devices generate more noise since CMOS output-voltage swings are larger, and in many cases faster, than those of advanced TTL devices. Thus, CMOS devices need more noise margin than TTL devices.

Complementary metal oxide semiconductor device input-output levels vary greatly with load conditions, and thus it is difficult to define generic noise margins (as can be done for TTL circuits). Input-output levels must be evaluated in each application to establish the actual noise margin.

4-2-1 CMOS input levels

Input levels of CMOS logic devices fall into two distinct categories: those with CMOS level inputs (AC, HC, etc.) and those with TTL-level inputs (ACT, HCT, etc.). Advanced CMOS devices with CMOS-level inputs (e.g., AC devices) have input thresholds at approximately 30 and 70 percent of actual device V_{cc}, as shown in Figure 4-5. Thresholds for HC devices are similar; the *high* threshold is the same, but the *low* threshold is at 20 percent. Thus, CMOS devices with CMOS input levels recognize input levels less than 30 percent (20 percent for HC) of V_{cc} as logic *low* inputs and those greater than 70 percent of V_{cc} as logic *high* inputs. Levels in the 30 to 70 percent range produce undefined responses.

In contrast to TTL thresholds, CMOS device threshold levels are a function of V_{cc} levels. Lowering or raising V_{cc} lowers or raises both *high* and *low* input thresholds. For standardization, most CMOS device input levels are specified with V_{cc} at 4.5 V. For example, AC parts are specified to have a minimum *high* input $V_{IH\ min}$ of 3.15 V, which is 70 percent of 4.5 V, and a maximum *low* input $V_{IL\ max}$ of 1.35 V, which is 30 percent of 4.5 V. Likewise, HC parts are specified to have a minimum *high* input $V_{IH\ min}$ of 3.15 V, which is 70 percent of 4.5 V, and a maximum *low* input $V_{IL\ max}$ of 0.9 V, which is 20 percent of 4.5 V.

Worst-case input-level limits for CMOS devices with TTL-input levels (e.g., ACT, FCT, and HCT) are the same as those of Schottky TTL devices (i.e., 0.8 V for *low* inputs and 2.0 V for *high* inputs).[9]

4-2-2 CMOS output levels

Complementary metal oxide semiconductor output levels vary greatly with load conditions, and to a lesser degree with temperature. Under static conditions CMOS outputs go to V_{cc} or ground when driving high-impedance loads such as CMOS inputs. However, CMOS output levels are generally not specified with ground or V_{cc} output levels, but there is little consistency in the load conditions under which they are specified. For example, in some data books, HC-device output levels (V_{OH} and V_{OL}) are specified for five output load conditions and three operating temperature ranges. Thus, it is difficult to pick a standard set of output levels for defining CMOS noise margins. However, for the purposes of this book, and in most CMOS application information, CMOS noise margins are based on the following output conditions:

1. A worst-case "true" CMOS *high* output is 4.4 V (0.1 V less than V_{cc} with V_{cc} at 4.5 V), and a worst-case CMOS *low* output is 0.1 V. Out-

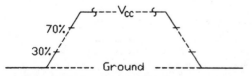

Figure   4-5 Advanced   CMOS-device   input thresholds.

put levels within 0.1 V of power or ground are only possible when devices are lightly loaded (50 μA or less), i.e., when driving high-impedance CMOS inputs.[8-10]

2. A worst-case TTL-compatible CMOS (ACT, HCT, etc.) minimum *high* output is defined as 3.7 V and a worst-case maximum *low* output as 0.4 V. In most cases, CMOS devices with TTL-compatible inputs have their output levels specified under load current conditions equivalent to those that might be encountered in similar TTL systems (i.e., HCT output levels are specified with loads equivalent to those used to specify LS output levels, and ACT output levels are specified with loads equivalent to those used to specify AS output levels).[8-10]

4-2-3 CMOS noise margins

Table 4-4 shows noise margins of several CMOS logic families with CMOS input levels and static CMOS output levels as defined above. Table 4-5 shows noise margins for several TTL-compatible families with TTL input levels and output levels based on sinking or sourcing current levels that might be encountered driving TTL devices.[11]

4-2-4 Summary of CMOS noise margin considerations

In general, CMOS devices with CMOS input thresholds have more noise margin than TTL devices. However, more noise margin is needed since the large voltage swings and fast edges of CMOS devices generate more noise than TTL devices. There are subtle system implications with respect to CMOS input thresholds that are often overlooked. In CMOS systems, threshold levels are a function of V_{cc}; however, in TTL systems thresholds are a function of voltage drops across diodes or base-emitter junctions that are referenced to ground. Since CMOS thresholds fluctuate with changes in V_{cc}, extra care must be taken where CMOS devices must interface with other devices (i.e., not

TABLE 4-4 Noise Margins of CMOS Logic Families with CMOS Input and Output Levels

Family	*Low*-level noise margin, V	*High*-level noise margin, V
HC	0.9	1.25
AC	1.25	1.25

TABLE 4-5   Noise Margins of CMOS Logic Families with TTL Input and
Output Levels

Family	Low-level noise margin, V	High-level noise margin, V
HCT	0.4	1.7
ACT	0.4	1.7
FCT	0.3	0.4

other CMOS devices) to ensure compatibility over worst-case V_{cc} levels.

Where CMOS devices with "true" CMOS input levels are used, the *high* and *low* noise margin specifications are the same. Thus, either signal polarity is acceptable for critical low duty cycle signals such as resets. However, for commonality with TTL systems, it is best to arrange critical low duty cycle signals so that they are in the *high* state most of the time since most systems will have a mix of TTL and CMOS devices.

High and *low* noise margin specifications are not the same for advanced CMOS devices with TTL input levels; *high* level signals have more noise margin than *low* level signals.* Thus, in those systems that use CMOS devices with TTL input levels, critical signals that have low duty cycles should be arranged so that they are in the high state most of the time. Only clocked devices that clock on low-to-high transitions should be used so that clock signals are *high* when system noise is at a maximum (noise is at maximum immediately following clock-switching edges).

Techniques for optimizing system noise immunity are covered in Part 2.

References

1. *FAST Applications Handbook 1987*, National Semiconductor Corp., South Portland, Me., 1988.
2. *FAST—Advanced Schottky TTL Logic Databook*, National Semiconductor Corp., Santa Clara, Calif., 1988.
3. *TTL Logic Data Book*, Texas Instruments Inc., Dallas, Tex., 1988.
4. *ALS/AS Logic Data Book*, Texas Instruments Inc., Dallas, Tex., 1986.
5. Heniford, William: "Muffling Noise in TTL," *The Electronic Engineer*, July 1969, pp. 63–69.

---

*When CMOS devices with TTL input levels are driven with TTL devices, signal worst-case noise margins are approximately the same as those of TTL systems and the same considerations apply with regard to optimum signal polarity for maximum noise rejection. When CMOS devices are used to drive CMOS devices with TTL inputs, *high*-level noise margin is greater than *low*-level noise margin by even a greater ratio than in TTL systems (see Tables 4-1 and 4-5).

6. Feulner, R. J.: "Solving Noise Problems in Digital Systems," *EEE, The Magazine of Circuit Design Engineering*, September 1967, pp. 79–83.
7. Lin, H. C.: *Integrated Electronics*, Holden-Day, San Francisco, Calif., 1967.
8. Funk, Dick: "Design Guidelines for CMOS Logic Systems," *Electronic Products*, March 28, 1984, pp. 75–79.
9. *SN54/74HCT CMOS Logic Family Application Report*, Texas Instruments Inc., Dallas, Tex., 1985.
10. *FACT—Advanced CMOS Logic Databook*, National Semiconductor Corp., Santa Clara, Calif., 1988.
11. *FCT—Fast, CMOS, TTL-Compatible Logic TECH NOTE*, Integrated Device Technology Inc., Santa Clara, Calif., 1986.

Sources of Transient Currents

The two major sources of transient currents in TTL and CMOS systems are internal feedthrough currents and load charging and discharging currents. Load charging and discharging currents are of particular concern when advanced TTL and CMOS logic devices are used. Their fast edges generate large transient currents in both the signal interconnection network and in the power and ground-reference distribution network. Unless proper high-frequency interconnection techniques are used, these transient switching currents will cause significant shifts in reference levels, large drops in local V_{cc} levels, and crosstalk or coupling into other signals, which all lead to various system upsets.

5-1 Transient Internal Switching Currents in TTL Devices

Each time TTL devices switch, internal transient currents flow between V_{cc} and ground. Most of this internal device transient switching current is the result of feedthrough in totem-pole output stages.[1,2] Feedthrough occurs as a result of both pull-up and pull-down totem-pole transistors being ON for a short period of time whenever an output switches. Device designers attempt to minimize feedthrough currents, but it is difficult to match the turn ON and OFF characteristics of transistors or the timing of separate drive signals. Advanced Schottky devices have added circuitry to help match the turn ON and OFF of output-stage transistors, but most circuits have some mismatch and some feedthrough current. Figure 5-1 shows a Schottky TTL totem-pole output stage and the equivalent circuit for a totem-pole output when both output transistors are ON at the same time. In the equivalent circuit (Figure 5-1b), the two 0.5-V voltage sources represent the two output transistors (ON V_{CE} for Schottky transistors is

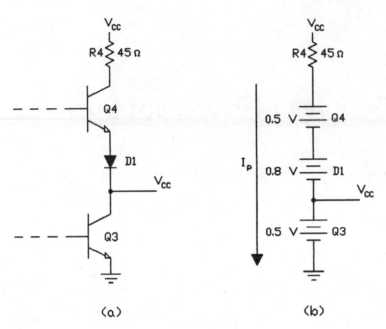

Figure 5-1 Totem-pole output and equivalent circuit during switching.

near 0.5 V) and the 0.7-V voltage source represents the forward drop across diode $D1$.

The upper limit for internal peak transient feedthrough current I_p is a function of the output-stage current-limiting pull-up resistor ($R4$ in Figure 5-1), the collector-to-emitter voltage drop V_{CE} of the two output transistors, and the forward voltage drop V_f across diode $D1$. That is,

$$I_p = \frac{V_{cc} - V_{CE(sat)} - V_f - V_{CE(sat)}}{R_{current\text{-}limiting}} \qquad (5\text{-}1)$$

If the value of the current-limiting resistor is 45 Ω (45-Ω pull-up resistors are the norm for standard FAST devices), the peak feedthrough current I_p could be as large as

$$I_p = \frac{5 \text{ V} - 0.5 \text{ V} - 0.8 \text{ V} - 0.5 \text{ V}}{45 \text{ } \Omega}$$

$$= \frac{5 \text{ V} - 1.8 \text{ V}}{45 \text{ } \Omega}$$

$$= 71 \text{ mA}$$

Actual peak feedthrough currents for FAST parts will be lower than 71 mA for several reasons. It is unlikely that both transistors will be fully ON during a significant portion of the switching period. Further-more, simple static transistor models, as used, are not sufficient for ac or transient analysis. In addition, inductance effects, which are not addressed in the equivalent circuit, tend to have a significant limiting effect at advanced Schottky switching frequencies. Observed currents tend to be more in the range of 10 to 30 mA for FAST TTL devices rather than 71 mA. Nonetheless, the equivalent circuit does illustrate the basic cause of internal feedthrough switching currents.

Feedthrough current duration is typically on the order of 1 to 2 ns for advanced Schottky devices. The fast switching speeds of advanced Schottky devices necessitate the close matching of ON-OFF character-istics. Devices that switch in 2 to 5 ns cannot have long overlap times. Some of the older, slower logic families have poor ON-OFF matching and long overlap times, on the order of 10 ns. As a result, they some-times have larger feedthrough currents than the more modern fami-lies. To limit feedthrough current in the older families IC designers used pull-up current-limiting resistors with larger values. For exam-ple, (standard) TTL devices use 130-Ω current-limiting resistors.

Output-stage transient feedthrough switching currents will vary with device, logic family, operating temperature, and load conditions. Thus, systems with poor power and ground distribution networks that operate correctly given one temperature and set of conditions may not operate given another temperature and another set of conditions. If conditions are changed, the transient feedthrough currents, which may cause different noise or reference-level disturbances, may change. If different parts are used, feedthrough current and operating margin may also change.

5-2   Transient Internal Switching Currents in CMOS Devices

Most CMOS logic devices are built with complementary stages that use p-channel FETs for pull-up switches and n-channel FETs for pull-down switches. Complementary stages have no dc current paths, other than leakage, between V_{cc} and ground. Yet, under dynamic condi-tions, complementary stages have the potential for significant inter-nal transient feedthrough switching currents; both pull-up and pull-down FETs may be ON for short, overlapping periods during switching.[3]

Figure 5-2a shows a simple CMOS inverter with one complemen-tary stage. During switching, both the pull-up FET (Q1) and the pull-down FET (Q2) may be ON for a short time. Hence, the switching transition equivalent circuit is two resistors between V_{cc} and ground as shown in Figure 5-2b. Gates and other CMOS logic devices have

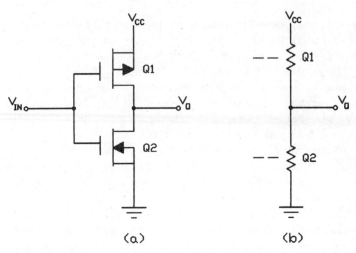

Figure 5-2 Inverting CMOS buffer and equivalent circuit during switching.

more complicated equivalent circuits than the inverter shown in Figure 5-2, but they have analogous dynamic current paths. Gate internal structures have more than one pair of FETs in series between V_{cc} and ground.

One approach for estimating feedthrough current for a simple inverting buffer is to divide the supply voltage V_{cc} by the sum of the ON impedances (resistances) of the two FETs. Typical ON resistance R_{ON} of advanced CMOS output-stage FETs is near 10 Ω; thus the peak internal transient switching current I_p could approach

$$I_p = \frac{V_{cc}}{2R_{ON}} = \frac{5\ V}{2(10\ \Omega)} = 0.25\ A \tag{5-2}$$

Actual transient feedthrough currents do not reach levels of 0.25 A. Both FETs are not fully ON at the same time, and most output-stage FETs go into current limit below 0.25 A. Advanced CMOS outputs with 24-mA worst-case static output drive ratings have a typical static output impedance of 10 Ω, but they current-limit at approximately 150 mA under typical conditions. Thus, a more exact CMOS complementary stage equivalent circuit (during switching) consists of two current sources in series (as shown in Figure 5-3). The pull-up current source I_p represents the pull-up FET in current limit and the pull-down current source I_n represents the pull-down FET in current limit.

The more exact equivalent circuit is also of limited use for predicting peak transient feedthrough current since the values of the current

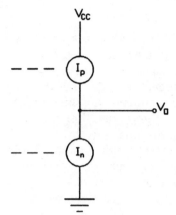

Figure 5-3 More exact switching transition equivalent circuit.

sources are unknown. The device output short-circuit current rating may be appropriate in certain cases, but in general, feedthrough current will not reach short-circuit current magnitudes. Both devices will not be fully ON at the same time. Peak transient feedthrough current magnitude is difficult to establish with simple models, although they do show the basic cause of feedthrough currents. When more accurate feedthrough current data are needed, most CMOS logic device manufacturers provide plots, such as Figure 5-4, that show feedthrough current versus input voltage. Feedthrough current I_{cc} in Figure 5-4 is normalized since the plot is not intended to represent a particular device. However, Figure 5-4 does show the typical profile of feedthrough current versus input voltage.

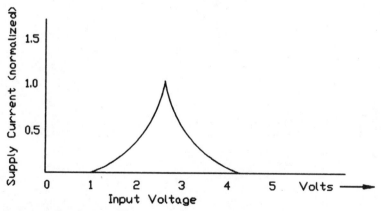

Figure 5-4 Supply current versus input voltage.

5-3   Transient Load Currents

Advanced TTL and CMOS devices with fast edge speeds and large voltage swings produce large transient load currents. Each time the signals change level, line and load capacitances must be charged or discharged, and the faster the rise or fall time, the larger the charging or discharging current (see Figure 5-5). Large charging currents increase the possibility that noise will upset the system. For example, large charging currents that encounter high-impedance points in either the signal, power, or ground interconnection systems may cause voltage transients in excess of logic device noise margins. Also, lines near lines with large transient currents may experience excessive cross coupling. The potential for system upset is greatest when advanced CMOS is used; CMOS output-voltage swings are larger than those of TTL, and as a result, CMOS transient switching currents are larger.

5-3-1   Charging current for a lumped capacitance load

When a load can be treated as a lumped load, i.e., the load is very near the source, peak charging currents can be estimated using RC-network calculations.[4,5] The charging current for the circuit in Figure 5-6 is

$$i = \frac{V_{cc}}{R_o} \, \epsilon^{-t/R_o C} \qquad (5\text{-}3)$$

and the peak transient current I_p at time t equals zero is

$$I_p = \frac{V_{cc}}{R_o} \qquad (5\text{-}4)$$

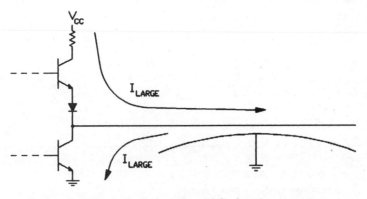

Figure 5-5   Capacitance load charging and discharging currents.

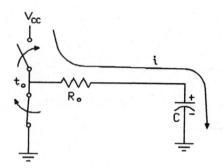

Figure 5-6 Circuit when charging a lumped capacitance load.

It is difficult to calculate charging currents to great accuracy because the driver output impedance (R_o in Figure 5-6) is usually not accurately known. However, peak current calculations using nominal values for output impedance are usually sufficient for most purposes.

5-3-2 Current-source charging currents

In most applications that use TTL or CMOS devices to drive transmission lines or large lumped capacitance loads, driver output transistors are in current limit for most of the switching transition interval. Thus, during signal transitions, both TTL and CMOS outputs should be treated as current sources as shown in Figure 5-7, and device output short-circuit current should be used to estimate peak output current. Output short-circuit current is listed on most data sheets. If the short-circuit current is not listed on the data sheet of interest, it may be found in the general logic family specifications.

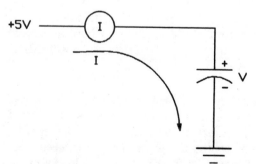

Figure 5-7 Current source charging a capacitor.

5-3-3   Charging current for a constant rate
of change of voltage

When the output rate of change dv/dt is known, the lumped capacitance C charging current I can be calculated to a good order of approximation using Equation (5-5).

The rate of change of voltage with time when a capacitor is charged with a constant current is

$$\frac{dv}{dt} = \frac{I}{C} \tag{5-5}$$

When Equation (5-5) is rearranged,

$$I = C\left(\frac{dv}{dt}\right) \approx C\left(\frac{\Delta v}{\Delta t}\right) \tag{5-6}$$

For a typical TTL level change of 3 V in 3 ns and a 50-pF load, the resulting transient load current I_L is

$$I_L = (50 \text{ pF})\left(\frac{3 \text{ V}}{3 \text{ ns}}\right) = 50 \text{ mA}$$

The 50 mA is a significant transient current. If 50 mA is multiplied by the number of signals that transition at a given time in a typical large digital system, a very large number results. However, much larger transient load currents are to be expected when CMOS devices are used. Many advanced CMOS devices have signal transitions of 5 V in 2 ns. When a signal transitions 5 V in 2 ns while driving a 50-pF load, the peak transient load current I_L is

$$I_L = (50 \text{ pF})\left(\frac{5 \text{ V}}{2 \text{ ns}}\right) = 125 \text{ mA}$$

The 125 mA is a significant transient load current. Having many signals with transient load currents of 125 mA, plus the additive effect of complementary stage feedthrough currents, offers the potential for significant amounts of current-generated noise in the power distribution and signal interconnection system. Thus, when advanced CMOS devices are used, great care must be taken to ensure low-impedance interconnection systems.

Load capacitance.   In many applications, load capacitance will exceed the standard 50-pF test load. Thus, transient load currents will exceed the values calculated above. The larger the transient load currents, the greater the demands placed on the power and signal interconnections. Thus, the magnitude of transient feedthrough and load capaci-

TABLE 5-1   Typical Device and Interconnection
Capacitance

Inputs, 5 pF

Outputs, 7 pF

Bidirectional ports, 15 pF

Printed circuit board traces, 2 to 5 pF/in

Wire-wrap wires, 1 to 2 pF/in

Welded-wire wires, 1 to 2 pF/in

---

tance charging currents must be factored into the design of the power
and reference distribution system and into the sizing and placement of
local decoupling capacitors (see Chapter 7).

Typical TTL or CMOS device input-output and typical interconnection capacitances are listed in Table 5-1. The data in Table 5-1 are
useful for quick estimates of load capacitance or as a guide when actual data are unavailable. However, actual specified device capacitance and interconnection capacitance should be determined and used
in transient load current calculations when they are available.

Most digital interconnections consist of several inches of "wiring"
that connects to a number of devices. Busses in particular connect to a
large number of loads. Thus, it follows from Table 5-1 that loads of 50
pF or more are to be expected on most signal lines and 100 pF or more
on bussed lines.

5-3-4   Transmission-line loads

When devices are driving transmission lines (see Section 11-4 for the
definition of a transmission line), transient load currents I_L are equal
to the magnitude of the signal change ΔV divided by the transmission-
line characteristic impedance Z_o [see Equation (5-7) and Figure 5-8)].

$$I_L = \frac{\Delta V}{Z_o} \tag{5-7}$$

5-4   Summary

Internal-device and load-related transient switching currents are a
major source of noise in advanced TTL and advanced CMOS systems.
Their presence and nature must be understood and the power distribution system designed to accommodate them, or there is little chance
of success. It is also essential that high-speed systems have low-
impedance power distribution systems and local decoupling capacitors

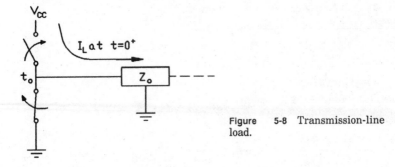

Figure   5-8   Transmission-line
load.

to replenish local transient-current demands. However, high-frequency interconnection techniques alone are not sufficient. Digital logic device interconnection schemes, such as synchronous design, that provide additional noise immunity are also required.[6]

Techniques for minimizing the detrimental effects of transient switching currents are described in Chapters 6, 7, and 8.

References

1. Mano, Morris M.: *Digital Design,* Prentice-Hall Inc., Englewood Cliffs, N. J., 1984.
2. Holt, Charles A.: *Electronic Circuits Digital and Analog,* Wiley, New York, 1978.
3. *Advanced CMOS Logic Designer's Handbook,* Texas Instruments Inc., Dallas, Tex., 1987.
4. Millman, Jacob: *Microelectronics, Digital and Analog Circuits and Systems,* McGraw-Hill, New York, 1979.
5. Millman, Jacob, and Herbert Taub: *Pulse, Digital and Switching Waveforms,* McGraw-Hill, New York, 1965.
6. Funk, Richard, and James Nadolski: "Advanced CMOS—Pinouts Are Not the Crucial Factor," *Electronic Engineering Times,* Monday, August 4, 1986.

Inductance and
Transient-Current Effects

Signal and power distribution system inductance is responsible for many of the difficulties encountered in high-speed systems. Yet, many designers continue to ignore interconnection inductance. The general impression seems to be that inductance and inductive effects are not significant in digital applications. Perhaps that was the case with the older, slower, logic families, but that is no longer the case. When advanced Schottky TTL or advanced CMOS logic devices are used, interconnection inductance is a major source of problems. The shortest possible connection has significant inductance and voltage loss when device switching transient currents have greater than 100-MHz frequency components.

6-1 Inductance

Inductance is a measure of the ability of a circuit to convert electromagnetic energy into a magnetic field. In certain applications, it is desirable to create a strong magnetic field, but this is not of importance in digital applications. Inductance must be minimized in digital systems to minimize signal energy loss (generating magnetic fields) and signal degradation. Yet, most texts that deal with the subject of inductance describe means for increasing inductance rather than means for decreasing inductance. They describe inductors that consist of numerous turns of wire and leave the reader with the general impression that a number of turns of wire are needed to achieve significant inductance. Even the symbol for an inductor, Figure 6-1, implies that a number of turns of wire are needed to create a useful inductor; that may be the case at 60 Hz but not at 100 MHz. Since voltage loss across an inductor is proportional to frequency, at 100 MHz a very small in-

Figure 6-1   Inductor.

ductor can cause significant loss. Thus, for the successful application of high-speed logic devices, power and signal interconnection inductance must be minimized.

Inductance can be minimized by good package design, printed circuit board design, and layout techniques (the shorter and closer to ground the better), and by employing special techniques for interconnection wiring. However, even when the best possible techniques are applied, transient voltage spikes due to inductive effects will be significant with respect to TTL or CMOS signal levels and noise margins. Thus, system design approaches, such as synchronous design practices, that minimize system susceptibility to transient voltage spikes must be used (see Chapter 8—Synchronous Design).

6-1-1   Physical and electrical factors that influence inductance

Self-inductance L, most often simply called inductance, of any current-carrying circuit is defined as the flux linkage per unit current.[1,2] That is,

$$L = \frac{\phi}{I} \tag{6-1}$$

where ϕ is magnetic flux and I is current.

Equation (6-1) is generally shown as

$$L = \frac{N\phi}{I} \tag{6-2}$$

where N is the number of turns of wire in an inductor. The more common equation for self-inductance, Equation (6-2), reinforces the idea that an inductor must be composed of numerous turns of conductor, but it should be noted that one (1) is an acceptable number for N in Equation (6-2).

Equations (6-1) and (6-2) show that inductance L is directly proportional to flux linkage. Since low inductance is desired for digital interconnections, it is important to understand what flux linkage is and how to minimize it. Flux density (flux linkage per unit area) is a measure of the ease with which flux can link with itself. The more concentrated the flux is, the greater the magnetic field and the greater the inductance. Flux does not need to link with other sources of flux, but flux, like current, must have a complete path. For inductance to be decreased, *the ease with which flux can link with itself* must be impeded. The two variables that determine *ease of linkage*, and thus the inductance, are the geometry of the circuit and the permeability of the magnetic medium. An iron core or other high-permeability µ material near a coil of wire increases the ease with which flux can link with itself and thus increases the inductance. Likewise, nearby low-permeability material reduces inductance. The permeability of the insulating material of printed circuit boards and of most insulating material is low; it is near that of air. Thus for conventional interconnection systems, little reduction in inductance can be accomplished by changing the material adjacent to signals.

However, digital designers have some control over the circuit geometry. Circuit inductance can be decreased by decreasing the length of the circuit (coil) or by arranging the circuit so that flux linkage is not enhanced (i.e., by arranging circuit topology so that flux cancels rather than enhances—e.g., coiling enhances flux linkage, so coiling should be avoided).

Equations for calculating the inductance of various physical arrangements of conductive elements can be found in numerous references. For many configurations the equations are quite complex. However, in all cases, flux linkage and inductance are proportional to the area enclosed by the circuit (current) path.[3,4] That is,

$$L \propto K(\text{area}) \qquad (6\text{-}3)$$

where K is a constant related to conductor geometry and the permeability of the surrounding material and *area* is the area enclosed by the current path. Since there is often little flexibility in the choice of insulating material permeability or conductor size, minimum inductance is achieved by minimizing the area enclosed by the current path. To minimize area, place signal lines and current return paths in close proximity. For example, even though their line lengths are the same, the current path in Figure 6-2a has less inductance than the current path in Figure 6-2b, since the enclosed area of 6-2a is less.

6-1-2 Transient voltage drop across inductors

The effect of a changing magnetic field $d\phi$ on an electric circuit is an induced voltage. The effect is described by Faraday's law, which states that the induced voltage v is

$$v = \frac{d\phi}{dt} \qquad (6\text{-}4)$$

for a single-turn inductor, which is the case for most digital circuits.

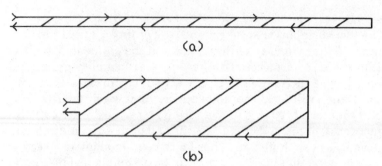

Figure 6-2 Inductance of current loops. (a) Low inductance; (b) high inductance.

However, the change in voltage across an inductor due to a changing current is of more interest to digital designers. Using Equations (6-1) and (6-4), it can be shown that a change of current i with time causes a corresponding change in the magnetic flux and induces a voltage v in the circuit.

First rearranging Equation (6-1) to

$$Li = \phi \qquad (6\text{-}5)$$

and differentiating both sides with respect to time t gives

$$L\frac{di}{dt} = \frac{d\phi}{dt} \qquad (6\text{-}6)$$

From Faraday's law the induced voltage is

$$v(t) = \frac{d\phi}{dt} \qquad (6\text{-}7)$$

and it follows that

$$v(t) = \frac{d\phi}{dt} = L\frac{di}{dt} \qquad (6\text{-}8)$$

and

$$v(t) = L\frac{di}{dt} \qquad (6\text{-}9)$$

is the familiar equation for the time-varying voltage across an inductor.

Since advanced Schottky and advanced CMOS devices produce power-supply current demands and signals with a large di/dt, it is imperative to minimize L in all current paths and to minimize undesirable transient voltages.

6-2   Inductance of Device Package Pins

There are situations where circuit inductance is much greater than desirable, yet there is little that can be done to reduce it. One such situation is the inductance of device package pins. Large dual-in-line (DIP) packages, in particular, may have relatively large pin inductances. Table 6-1 shows the range of typical pin inductances for several common packages.[5,6]

The two values listed in Table 6-1 are representative of the inductance of the longer and the shorter signal paths (pins) in or out of a given package style. The 20-pin plastic leaded chip carriers (PLCCs) are nearly symmetrical. Hence, there is little difference in the length, or inductance, of the pins (see last entry in Table 6-1).

To reduce package lead inductance, some manufacturers of advanced CMOS parts have changed from the more common end locations for power and ground on dual-in-line packages to double center power and ground pins to reduce inductance.[7] Center power and ground pins lower average and worst-case inductance between internal points on a chip and external ground or supply-voltage levels since average and maximum distances to external connections are reduced.

6-3   Ground Bounce

Ground bounce and power-supply droop are of great concern when advanced Schottky or advanced CMOS devices are used.[8] Their fast edges cause large, high-frequency transient load currents which also must flow in the ground and power leads of the switching device. Large transient currents in package leads in turn cause large shifts in chip reference and power-supply levels.

TABLE 6-1   Package-Pin Inductance

Package	Self-inductance, nH	
	Upper value	Lower value
14-pin DIP	10	3
20-pin DIP	14	3
20-pin PLCC	5	4

"GROUND BOUNCE" IS THE TERM USED TO DESCRIBE
TRANSIENT CHIP REFERENCE SHIFTS CAUSED BY PACKAGE-
PIN INDUCTANCE AND TRANSIENT-CURRENT FLOW TO
GROUND.
"POWER-SUPPLY DROOP" IS THE TERM USED WHEN
POWER LEAD INDUCTANCE AND TRANSIENT-CURRENT
FLOW FROM THE SUPPLY TO THE CHIP CAUSE A SHIFT IN
CHIP SUPPLY VOLTAGE.

Ground bounce (or power-supply droop) may upset device operation
or cause system logic errors. Ground-bounce (or power-supply-droop)
problems are compounded when several heavily loaded drivers in a
common package switch simultaneously.[9] One observable manifesta-
tion of ground bounce or power-supply droop is the appearance of transient voltage spikes on stable outputs when other heavily loaded devices in the (same) package switch.[10]

Figure 6-3 shows a common octal driver application where ground-
bounce spikes will be observable on the one stable output (power-
supply droop will occur for the opposite switching conditions). Ground-
bounce spikes will exist on the stable output when the other drivers
switch from a *high* to a *low* level. *High*-to-*low* switching causes a large
transient current to flow from the loads to ground through the package ground pin. The inductance of the pin and the large transient current causes a transient voltage drop (spike) across the ground pin. The
transient voltage spike causes the reference level of the internal chip
to shift in the positive direction, causing a positive voltage spike on
the stable output as shown in Figure 6-3. The magnitude of the spike
depends on the total load of the switched outputs. In many common
situations, the magnitude of the spike will exceed the noise margin of
the receiving device.

To find the magnitude of a transient ground-bounce voltage spike
for a circuit, such as shown in Figure 6-3, first the peak transient current must be determined. In the application shown in Figure 6-3,
seven buffers are simultaneously switching seven 50-Ω lines. Assuming that the level transitions are 3 V, the change in current is

$$\Delta I = 7\left(\frac{\Delta V}{Z_o}\right)$$

$$= 7\frac{3\text{ V}}{50\text{ }\Omega} \tag{6-10}$$

$$= 420\text{ mA}$$

Once ΔI is known, transient ground-bounce voltage is calculated from
Equation (6-9).

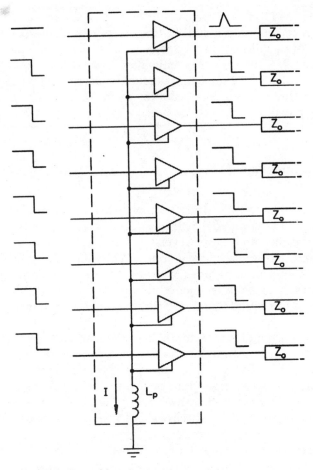

Figure 6-3 Ground bounce.

$$v(t) = L\left(\frac{di}{dt}\right)$$

which is approximated by

$$\Delta V = L\left(\frac{\Delta I}{\Delta t}\right) \tag{6-11}$$

For a ΔI of 420 mA, a signal transition of 3 ns, and a ground pin inductance of 10 nH, the ground-bounce spike ΔV is

$$\Delta V = (10 \text{ nH})\left(\frac{420 \text{ mA}}{3 \text{ ns}}\right)$$

$$= 1.4 \text{ V}$$

A value of 10 nH is used in the above calculations as representative of the inductance of a typical 20-pin dual-in-line package pin (see Table 6-1). A ground-bounce spike of 1.4 V greatly exceeds the TTL *low-level* noise margin (0.3 V); yet, 50-Ω dynamic loads are not unusual. In many cases, much lower board impedances must be driven; printed circuit (pc) boards may have effective impedances Z_o that are as low as 20 or 30 Ω. Large lumped capacitance loads have very low dynamic impedances.

6-4   Inductance of Component Power Connections

Device reference- or supply-voltage shifts are not limited to package lead inductance; imperfect power and reference distribution systems also degrade local reference and power levels. To minimize degradation of voltage and reference levels on circuit boards and motherboards, use low-impedance planes or grids to distribute power and ground. It is difficult to establish exact numerical values for the inductance of a voltage or ground plane. Tightly coupled (i.e., in close physical proximity for maximum distributed capacitance) planes or finely spaced grid systems exhibit very low effective impedance, 1 nH or less. However, to take full advantage of a low-impedance distribution system, the connections between the component power pins and the planes must be direct—no wiring or long pc board tracks. When pc boards are used, direct connections to power and ground planes are normal. However, when wire-wrap or stitch-weld universal boards are used, as is often the case during the developmental phase of a project, there is often a temptation to wire ground and power connections. Quick calculations can show the inappropriateness of wired power or ground connections.

On wire-wrap boards, it is difficult to achieve connections that are much shorter than 1 in. Various references show that the inductance of 1 in of number 30 wire is in the 20 to 50 nH range, depending upon how close the wire is to a ground plane or other return path.[11–13] The transient voltage drop ΔV across a 1-in wire with an inductance of 20 nH, and the same load and switching conditions as used for the component pin calculations above (seven outputs switching in 3 ns with 3-V level changes and 50-Ω loads), is given by Equation (6-11).

$$\Delta V = L\left(\frac{\Delta I}{\Delta t}\right)$$

$$= (20 \text{ nH})\left(\frac{420 \text{ mA}}{3 \text{ ns}}\right)$$

$$= 2.8 \text{ V}$$

A 2.8-V ground-wire spike plus a 1.4-V spike due to the inductance of the package pins is intolerable. A 3- to 4-V ground-bounce spike is much in excess of TTL or CMOS noise margins. Thus, the inductance of wired power or ground connections cannot be tolerated when high-speed logic components are used. However, even when all external connections are ideal, i.e., zero-impedance, package pin inductance can cause significant ground bounce. Thus, logic techniques must be devised to negate the possible detrimental effects of ground-bounce voltage spikes.[14] Synchronous design practices (Chapter 8) are one means of limiting the possible detrimental effects of ground bounce. Synchronous design prohibits the use of asynchronous inputs on storage elements (flip-flops) or cross-coupled gates for performing operational logic functions. Thus, short spikes, such as might be generated by ground bounce, cannot directly upset syncronous systems.

To further limit the possible detrimental effects of ground bounce or power-supply droop, the buffers used to drive critical signals, such as clocks and signals that go to asynchronous inputs, such as resets and presets, must be segregated into separate packages. The purpose is to prevent the ground bounce that results from heavily loaded buffers from upsetting other buffers (in a common package). For example, different frequency clock signals should not use buffers in a common package. All the buffers in an octal package should not be used if all the buffers are driving heavily loaded lines and all the signals could switch at once; clock signal fan-out buffers are an example. When the number of buffers used in an octal package is limited because of ground-bounce considerations, the buffers nearest the ground pin should be used.

6-5   Conclusion

When applying modern high-speed TTL and CMOS devices, designers must minimize power and signal interconnection inductance. For minimum inductance, signal and return current paths must be in close proximity and as short as possible. Low-inductance interconnections are needed to ensure that switching transients generated by active devices driving large capacitance loads will not upset other devices in the system. The inductance and the capacitance of "real world" interconnections cannot be ignored if a reliable high-speed system is to be designed.

6-6   Summary of Techniques for Minimizing Inductance and Transient Current Effects

> For minimum inductance, signal and return current paths must be in close proximity and as short as possible.

Techniques for minimizing the inductance and the detrimental transient-current effects in power and signal distribution systems include:

1. Power and ground planes should be used to distribute power and ground. Ground reference planes are essential for high-speed systems. Power may be distributed with a ladder or grid configuration but never with point-to-point wiring.

2. Continuous ground planes with no avoidance areas in IC package or connector pin fields are required on circuit boards and motherboards.

3. Decoupling capacitors must be located as close as possible to active circuitry (i.e., components). A good rule of thumb is one adjacent decoupling capacitor per high current driver or memory device and one for every two packs of random logic (see Section 7-4-1).

4. Power supplies should be located as close as possible to the powered circuitry, and to minimize the area between the lines and hence the inductance, use twisted-pair lines to bring power to local power planes when the power supply cannot be connected directly to the power and ground planes. When the power supply cannot be directly connected to the motherboard planes, local bulk decoupling may be needed at the power entry point (see Section 7-4-2).

5. Ground pins must be evenly distributed across all connector-pin fields, including custom-chip or multiple-chip packages. The number of connector ground pins required to minimize the effects of transient load currents must be determined from the system ground upset error budget. A rule of thumb for the number of connector ground pins is a minimum of one ground pin per inch of connector length, or no less than one ground pin per eight outputs.

6. Signal lines must run near ground (reference) planes so that the area between a signal and its current return path is minimized. Minimum area minimizes the inductance of the complete current path. When signals cannot be routed near a ground plane (such as between separate units), signals should be sent via a differential twisted-pair line (the need for differential signals between units not referenced to a common ground plane is described in Chapter 10).

7. When using prototyping boards (such as wire wrap, stitch weld, etc.), the power and ground planes must be continuous through the package- and connector-pin fields and package (or socket) ground and power pins must be soldered directly to the planes on pc boards and soldered directly to the planes using solder washers or clips on

universal boards. Wiring of power and ground connections with discrete wires is not permissible because they cause excessive inductance.

References

1. Fitzgerald, A. H., and D. E. Higginbotham: *Basic Electrical Engineering*, McGraw-Hill, New York, 1957.
2. Corcoran, George F., and Henry R. Reed: *Introductory Electrical Engineering*, Wiley, New York, 1957.
3. Langford-Smith, F.: *Radiotron Designer's Handbook*, Radio Corporation of America, Harrison, N. J., 1960.
4. Lee, Reuben, Leo Wilson, and Charles E. Carter: *Electronic Transformers and Circuits*, Wiley, New York, 1988.
5. *Advanced CMOS Logic Design Considerations*, SCLA004, Texas Instruments Inc., Dallas, Tex., 1986.
6. *FAST Applications Handbook 1987*, National Semiconductor Corp., South Portland, Me., 1988.
7. *Advanced CMOS Logic Designer's Handbook*, Texas Instruments Inc., Dallas, Tex., 1987.
8. Shear, David: "EDN's Advanced CMOS Logic Ground-Bounce Tests," *EDN*, March 2, 1989, pp. 88–97.
9. *Simultaneous Switching Evaluation and Testing*, Texas Instruments Inc., Dallas, Tex., 1987.
10. *FCT—Fast, CMOS, TTL - Compatible Logic Tech Note*, Integrated Device Technology Inc., Santa Clara, Calif., December 1986.
11. Morrison, Ralph: *Grounding and Shielding Techniques in Instrumentation*, Wiley, New York, 1967.
12. Harper, Charles A.: *Handbook of Electronic Packaging*, McGraw-Hill, New York, 1969.
13. Mohr, Richard J.: "Interference Coupling—Attack It Early," *EDN*, July 1, 1969, pp. 33–41.
14. Funk, Richard, and Nadolski, James: "Advanced CMOS—Pinouts Are Not the Crucial Factor," *Electronic Engineering Times*, Monday, August 4, 1986, p. 33.

System Design Guidelines

Power Distribution

Advanced Schottky and advanced CMOS systems require the use of high-frequency techniques throughout the power distribution system. The power interconnection system must have the ability to supply large transient load and device currents, plus the ability to prevent noise generated by local load and device transient switching currents from propagating to other parts of the system. Of primary importance is the use of very low-impedance interconnections throughout the distribution systems and the use of decoupling capacitors located very close to all high-speed digital devices.[1] Meeting the dc and transient-current demands of large, high-current high-speed digital systems is a difficult electrical and mechanical design task. The challenge is to deliver and maintain the proper voltage levels to the board-level components during all worst-case transient switching conditions. Yet, the power and return current must traverse a complicated path of wires, connectors, circuit board, and motherboard power and ground planes. Even when the best design practices are followed, the current path is plagued with resistance and inductance that will degrade the quality of the delivered power. The task of the power interconnection system designer is to ensure that the degradation does not exceed the operating limits of the logic devices being used.

To ensure that supply-voltage levels are not degraded beyond the limits of the logic components, the power distribution system designer must consider possible losses in the following areas:

1. Component power and ground connections to board power and ground planes
2. Circuit board power and ground planes
3. Circuit board to motherboard power and ground connections
4. Motherboard power and ground planes

5. Motherboard power and ground plane connections to power and ground feeder lines or busses

6. Power and ground lines, or busses, between motherboard and power supplies

7. Power and ground feeder-line connections at power supplies

Techniques for determining, controlling, and compensating for losses in each of the above listed areas are described in the following sections.

7-1 Component Power Connections

Connections between component power pins and circuit board power and ground planes must be as direct as possible to minimize inductance.[2] Direct connections to planes are seldom difficult to achieve when printed circuit boards are used. However, when wire-wrap or stitch-weld boards are used, as is often the case during the developmental phase of a project, there is often a temptation to use wire for power and ground connections. Quick calculations show the inappropriateness of wire connections.

For a wire-wrap board it is difficult to achieve wired connections of much less than 1 in. Various references show the inductance of a 1-in length of number 30 wire to be in the 20- to 50-nH range, depending on how close the wire is to a ground plane or current return path (see Chapter 6).

The transient voltage spike across a 1-in wire resulting from a 2.5-V transition ΔV with a 50-Ω load (Z_o) is significant with respect to TTL noise margin as shown below.

When an output switches, the change in current ΔI is

$$\Delta I = \frac{\Delta V}{Z_o}$$

$$= \frac{2.5\ \text{V}}{50\ \Omega} \tag{7-1}$$

$$= 50\ \text{mA}$$

The resulting supply-voltage droop or ground bounce can be obtained from Equation (6-11) using the lower value for the inductance of a 1-in wire, 20 nH, and a 2-ns transition

$$\Delta V = L\left(\frac{\Delta I}{\Delta t}\right)$$

$$= (20 \text{ nH})\left(\frac{50 \text{ mA}}{2 \text{ ns}}\right)$$

$$= 0.5 \text{ V}$$

A 0.5-V spike may not upset the receiving devices, depending upon the response of the receiving devices, but a 0.5-V spike does exceed the TTL *low* noise margin. Furthermore, the above set of circumstances are very benign. Only the current resulting from a change in *one* output is addressed; in most cases more than one output in a package will change at the same time, and the transient voltage drop due to the inductance of the package pins and the loss in the local power or ground distribution system (if planes are not used) must be added to the value calculated above. Any one source of noise should not exceed the noise margin of the devices used (0.5 V for *high* TTL inputs, 0.3 V for *low* TTL inputs). Some reserve is needed for other sources of noise, such as crosstalk and dc reference offsets. A system noise budget should be established to allocate limits for each major source of noise.[3] The total error budget must not exceed the noise margin of the devices used.

Ground bounce and power-supply droop in excess of TTL or CMOS noise margins is unavoidable when advanced Schottky or advanced CMOS devices are used, even when the best electrical design rules are followed. Thus, logic techniques must be devised to negate the possible detrimental effects of ground bounce or power-supply droop spikes. Synchronous design practices (which are described in Chapter 8) are one means of limiting the effects of ground bounce. Synchronous design prohibits the use of asynchronous inputs on storage elements (flip-flops), or cross-coupled gates for logic operations. Thus, short spikes, such as might be generated by ground bounce, cannot directly upset logic devices.

To further limit the possible detrimental effects of ground bounce or power-supply droop, drivers and gates used to transmit certain critical signals, such as clocks and initialization signals (dc resets and presets, or other asynchronous inputs), must not share packages with other signals. Buffers and gates used to drive signals to sensitive input (for example, clock inputs) must be segregated into separate packages so that ground bounce (from a heavily loaded signals) will not generate spikes on sensitive signal lines. For example, clock signals of different frequencies should not use drivers in a common package; clock signals of the same frequency and phase can use multiple drivers in a common package. However, even when the signal frequencies and phases are the same, all of the buffers in an octal driver package should not be used in critical clock-driver applications to limit the transient current. The buffers used should be those nearest the package ground pin.

7-2 Voltage Drop across a Plane

To minimize local transient-voltage loss, circuit-board power and ground must be distributed by low-impedance planes or grids. The drop across a circuit board or motherboard plane is difficult to calculate since currents in planes are not confined to known paths. One useful approach for estimating dc voltage drops across a plane is to segment the plane into squares, as shown in Figure 7-1, and then sum the more manageable and more easily determined drops across the individual squares. First, determine the current requirements of each square (i.e., the supply-current requirements of components or circuit board connector pins that are located in that portion of the plane). Second, average the current for each row of squares and make the assumption that the current density is uniform in each square in a given row across the plane (i.e., from left to right in Figure 7-1). Third, determine the cumulative current in each square in a column of the squares working from the top of the plane to the bottom. Use the sheet resistance of the plane and the current in a given square to calculate the IR drop across each square. Sum the drops across the squares to arrive at the drop across the plane or other points of interest.

Resistance of a section of a plane. For a section of a plane, with dimensions as shown in Figure 7-2, the resistance is

$$R = \rho\left(\frac{l}{zw}\right) = \rho\,\frac{l}{\text{area of the cross section}} \qquad (7\text{-}2)$$

where ρ is the resistivity of the plane. For a square section ($l = z$), the resistance is

$$R = \frac{\rho}{w(\text{thickness})} = \rho_s(\text{sheet resistance}) \qquad (7\text{-}3)$$

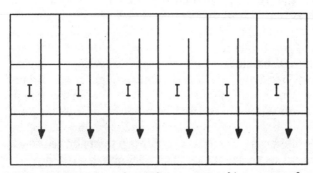

Figure 7-1 Ground or voltage plane segmented into squares for IR loss calculations.

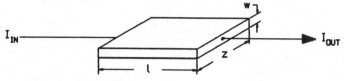

Figure 7-2 Section of a plane.

The approximate sheet resistance for three common weights of copper printed circuit board conductive layers is:[4]

0.5-oz copper plane $\approx$ 1.0 mΩ per square

1.0-oz copper plane $\approx$ 0.5 mΩ per square

2.0-oz copper plane $\approx$ 0.25 mΩ per square

Most multilayer printed backplane and circuit boards are built with 1-oz copper planes. The 2-oz planes are used in very high current applications. The 0.5-oz material is seldom used for backpanel or motherboard power and ground planes, but it is used for signal layers and in a few cases for low-current circuit board planes.

The sheet resistance used to calculate the drop across a plane should be derated from the solid plane, the amount being based on the number and location of the voids. In many applications, voids resulting from clearance holes for vias and package or connector pins will result in cross-sectional area losses in excess of 50 percent. Cross-sectional loss must be determined, and the sheet resistance should be adjusted as appropriate.

Voltage loss in planes that is due to inductance as well as resistance must be considered, but it is difficult to establish exact numerical values for the inductance of a voltage or ground plane. However, tightly coupled (in close physical proximity for maximum distributed capacitance) planes or finely spaced grid systems exhibit very low effective impedance—1 nH or less. Thus, circuit board and motherboard power and ground must be distributed by low-impedance planes or grids to minimize transient voltage loss.

7-3 Power Supply–to–Motherboard
Connections

Advanced Schottky and advanced CMOS device systems require power sources with low-impedance outputs and power supply–to–motherboard interconnecting conductors and connections with low impedance to meet the high dc current and high pulse-current requirements.[5] It is essential to minimize inductance L in order to control undesirable transient voltage drops. Minimum inductance is achieved by minimizing the area enclosed by the current path, i.e., keeping the current path as short as possible. In low-current applications, the dc resistance of the power connections is seldom a problem; in high-current applications (greater than 50 A), the dc resistance of the power connections can be a major problem. The designer should make every effort to minimize the length of the V_{cc} and ground conductors between the power supply and the power and ground planes in backpanels or motherboards. The optimum arrangement is a dedicated power supply located as close as possible to each motherboard. A local power supply for each unit keeps the V_{cc} and ground connections as short as possible and minimizes the occurrence of circulating ground-loop currents.

7-3-1   Remote power-supply connections

If it is not possible to directly connect the power supply to the power and ground planes in the motherboard, the connections should be made with twisted pair lines or low-impedance closely coupled bus bars.[6] Minimizing the inductance of remote power connections is as important as minimizing the resistance due to the pulsed current demands of advanced Schottky or advanced CMOS systems. To help supply peak-current demands, locate some local bulk capacitance near the power connection to the motherboard. Local bulk capacitors also help reduce radiated noise. In remote applications with very high transient-current demands, motherboards should have bulk capacitors distributed around the perimeter. The value of local bulk capacitance should be in the range of 50 to 100 times the value of the worst-case simultaneously switched load capacitance (see Chapter 5). Bulk capacitors must be chosen carefully. Many large value capacitors are sensitive to surge current, ripple current, reverse voltage, and overvoltage (see Section 7-4-2). The best solution is to design the power distribution system so that bulk decoupling capacitors are not needed. A local power supply with short low-impedance connections to low-impedance power planes achieves that purpose. However, in those cases where power supplies cannot be located near the load, they should be connected as shown in Figure 7-3.

When the power supply cannot be located near the load, overshoot at turn ON and undershoot at turn OFF are common occurrences. Power-supply overshoot and undershoot are enhanced by the amount of inductance in the supply path. Thus, the inductance of the power conductors must be minimized to control overshoot and undershoot, as well as to ensure a low-impedance path for transient load demands. The possibility of overshooting at turn ON, due to inductance between

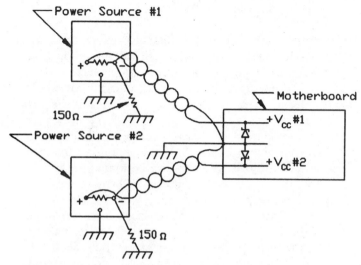

Figure 7-3 Remote power-supply connections.

the supply and the load, can be minimized by mounting a 10- to 100-Ω resistor directly across the power-supply terminals to provide a real (noninductive) load at the power-supply terminals. A resistor at the source has additional benefits: It tends to reduce reflected noise by matching (terminating) the power lines. A power supply may have very low dc output impedance and be capable of sourcing and sinking large amounts of dc current, yet the output may look like a high impedance to an incident ac noise waveform (see Chapter 11). If an output has a high ac impedance, noise pulses may require numerous trips between the supply and the load for the energy to be dissipated. It is always desirable to dissipate noise energy quickly so as to minimize the overall system noise.

Power-supply return or reference outputs should not be connected to ground at the supply itself. The power supply ground lines, backpanels (or motherboards), ground planes, chassis, racks, etc., should be connected to earth or chassis ground at one central point only.[7] In most applications, the best location for the central ground point is where the ground lines connect to the motherboard or backpanel. A single system ground point prevents circulating ground currents. Circulating ground currents introduce noise that can upset system operation; thus extraneous ground currents must be avoided.[8] To ensure that power supplies have a reference at all times, connect the remote power-supply return or reference output to earth ground through a resistor, such as 150 Ω, as shown in Figure 7-3. A resistor prevents ground-loop currents yet provides a reference so that a supply does not go to some harmful level if the line to the central ground point is disconnected.

Systems or units must not be allowed to float, to reduce the possibility of unsafe voltage levels. All systems must have a solid, direct connection to earth ground to protect personnel from electric shock and to prevent damage to interface devices.[9,10] If a unit is allowed to float, input-output devices may not operate correctly, even if they are not damaged.

The use of a single power source for multiple units, e.g., backpanels, motherboards, chassis, etc., is undesirable. When a single unit feeds multiple units, the process of grounding the individual units creates multiple grounds and introduces the possibility of large circulating ground currents. When multiple units must be powered from a single source, the single-point earth ground must be located at the power source as shown in Figure 7-4. Figure 7-4 shows the correct way to connect multiple units to a single source. Separate lines should be used to connect each unit to the supply. Units should not be serially connected so that unit-to-unit interference is limited. In most situations, a supply for each unit, which allows each unit to be directly grounded, is a more desirable configuration.

7-3-2 Voltage loss in power conductors

Many of today's digital systems have power-supply current requirements of 100 A or more. For applications that require more than 100

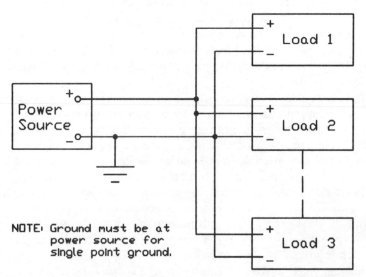

Figure 7-4   Single source to multiple load power and ground connection.

A, power conductors must be carefully designed to minimize voltage loss. At more than 100 A, very small resistances are significant. Even in low-current applications, power-conductor loss is often underestimated. In all applications, dc voltage loss in power wiring or bus bars must be evaluated. Power conductors must be sized so that the loss in the power leads, when added to the voltage tolerance of the power supply, plus an allowance for noise and other losses does not result in a voltage at the components that is below their minimum specified operating level. The minimum operating level for most TTL and CMOS devices is 4.5 V for military temperature range rated devices and 4.75 V for commercial rated devices. When calculating power conductor loss, be sure to consider both power and ground return lines.

When calculating power conductor loss, the manufacturer's resistance specifications should be consulted for actual resistance data. However, if stranded copper wire is used, the voltage drop can be estimated using the data listed in Table 7-1. Table 7-1 lists the approximate resistance (in milliohms per foot) for several common sizes of stranded copper conductor.[11,12]

7-3-3 Voltage loss in power connections

Connector and power line junction losses are a serious problem in high-power systems. Large high-speed digital processing systems often require supply currents in the 100- to 200-A range. In such applications, a small connector or junction resistance can result in large voltage losses. To minimize the voltage loss, the power connections must be tight and clean and of sufficient contact area (to limit the cur-

TABLE 7-1   Resistance of Stranded Copper Conductor

AWG*	mΩ/ft
4	0.27
6	0.43
8	0.67
10	1.2
12	1.8
14	2.9
16	4.5
18	5.9
20	9.5
22	15
24	24
26	40
28	64
30	100

*AWG, American Wire Gage.

rent density).[13] In most large current applications (greater than 100 A), it is not practical to use a quick-disconnect power connector; bolted junctions are required.

Determining the resistance of power connections under all conditions is a difficult task. Contact pressure, oxidation of mating surfaces, and actual mating surface area (there may be high or low spots) are all critical parameters that are difficult to establish initially, and it is difficult to predict their values over time. To compensate for these uncertainties, power connections should be designed with a great deal of safety margin.

Multiple power and ground connections to backpanel planes (backplanes) may be required in high-current applications in order to limit the current density and voltage loss in backplanes in the vicinity of power connections. When multilayer printed circuit motherboards with thin planes are used in high-current applications, the current density in the vicinity of power connections is of special concern. In such applications, the voltage loss in the plane in the vicinity of the connections must be evaluated and corrective measures (such as multiple connections) taken if the loss is significant.

Radial resistance of a thin plane from a point.   The radial change in resistance dR of a thin plane as a function of distance r from a point or circular void (see Figure 7-5) is[14]

$$dR = \rho_s \frac{dr}{2\pi r} \tag{7-4}$$

where ρ_s is sheet resistance and dR, the change in resistance between two points (r_1 and r_2) in a thin plane, is

$$dR = \frac{\rho_s}{2\pi} \int_{r_1}^{r_2} \frac{dr}{r} \tag{7-5}$$

and the resistance R between r_1 and r_2 is

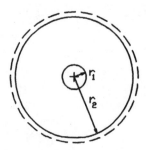

Figure 7-5   Variables used to determine the radial resistance of a plane.

$$R = \frac{\rho_s}{2\pi} \ln \frac{r_2}{r_1} \qquad (7\text{-}6)$$

When multilayer printed circuit motherboards with thin planes are used in high-current applications, Equation (7-6) should be used to calculate the resistance in the vicinity of the power and ground connection, and the results should be used to calculate possible voltage loss.

Example calculations.   The voltage drop in the vicinity of a 10-A and a 200-A connection to a 1-oz copper plane is determined to illustrate the possible voltage loss at power or ground plane to power conductor junctions. A 10-A connection is typical of circuit board current requirements. Many large systems have motherboards with requirements of greater than 200 A.[15]

For the 10-A case, a 0.030-in-diameter hole is used as a starting point. A 0.030-in-diameter hole (r_1 = 0.015 in) is consistent with the mounting needs of a 10-A connector. Using Equation (7-6), the radial resistance of a 1-oz copper plane starting from a 0.030-in-diameter hole is shown in Figure 7-6. At a distance of 2 in, the resistance is approximately 0.4 mΩ.

The voltage loss V_{loss} across the 2 in of plane is

$$V_{\text{loss}} = (10 \text{ A})(0.4 \times 10^{-3} \text{ }\Omega)$$

$$= 4 \text{ mV}$$

The 4 mV is not a significant loss. Thus, similar low-current connections to 1-oz copper planes should not cause concern in most applications.

A 0.375-in-diameter hole is typical of the bolt or stud size that

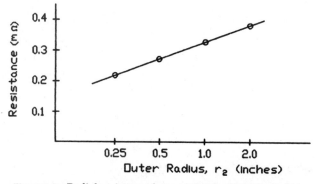

Figure 7-6   Radial resistance from a 0.030-in-diameter hole in a 1-oz copper plane.

might be used to connect a 200-A power-conductor terminal to a motherboard. The radial resistance of a 1-oz plane starting from a 0.375-in-diameter hole is shown in Figure 7-7. Starting from a 0.375-in-diameter hole, the radial resistance is about 0.2 mΩ at a radius of 2 in, and the voltage loss V_{loss} with 200 A of current flow is

$$V_{\text{loss}} = (200 \text{ A})(0.2 \times 10^{-3} \ \Omega)$$

$$= 40 \text{ mV}$$

A 40-mV loss at both the power and ground connections to a motherboard must be evaluated in conjunction with other possible voltage losses in the power system. Such a large loss, coupled with other unavoidable junction and line losses in the power supply–to–motherboard power feeder system, may adversely affect system operating margin. Where high-current connections must be made to a plane, connection losses can be minimized by connecting the power and ground conductors to the motherboard at multiple points.

Connections between dissimilar metals. Direct connections between current-carrying conductors of dissimilar metals, such as aluminum backplanes and copper bus bars or wires, must be avoided. Untreated aluminum quickly forms a thin oxide film, which causes a high-impedance joint. Power connections to aluminum backplanes must be made using bimetallic transition joints, or contact surfaces must be coated with an appropriate joint compound.[16,17] Joint compounds contain chemicals that dissolve surface oxide films and seal joints against moisture and further oxidation. A joint compound, such as Alcoa No. 2 Electrical Joint Compound, or an equivalent compound, should be used on all untreated aluminum-to-aluminum or aluminum-to-copper electrical connections. Transition joints are formed by creating a metallic bond between an aluminum backplane and some other metal,

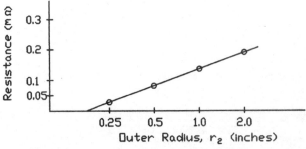

Figure 7-7   Radial resistance from a 0.375-in-diameter hole in a 1-oz copper plane.

such as copper, that does not form a high-impedance oxide film. The metallic bond between the aluminum and copper prevents an oxide layer from forming at the aluminum-to-copper junction, and the copper outer layer provides a low-impedance surface for power connections.[18]

7-3-4   Overvoltage protection

Power-supply overshoots at turn ON or failures that cause a higher than normal output level must be guarded against.[19,20] For long-term system reliability, power supplies must be limited to levels that will not instantly destroy devices or stress them so that their useful life is reduced. The goal is to keep all steady-state and transient supply-voltage levels below 7 V, which is the absolute maximum rating for most advanced TTL and CMOS logic components. It is not uncommon for power supplies to fail in such a manner that the output voltage goes to a higher than normal level. Yet, it is usually carelessness in the adjustment or connection of the supply to the load, or overshoot at turn ON, that causes an overvoltage problem. Whatever the reason, it is very discouraging to have an entire system destroyed because of an overvoltage condition on the power supply (and it most assuredly will be a very unpleasant task to explain the situation to your supervisors).

Thus, to ensure that supply voltages do not exceed device limits, equip all power supplies with overvoltage shut-down circuits or crowbars. In addition, all systems should have a passive clamping device (one not dependent on some other supply to operate) located directly and permanently at the load (so that it will not be neglected or improperly connected). In most applications, passive clamps should be located where power is connected to motherboards or backpanels as shown in Figure 7-8. A power-transient suppressor, such as a 1N5808 or 1N5809 or some equivalent device, is recommended. The passive

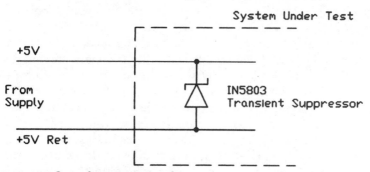

Figure 7-8   Overvoltage protection clamp.

clamping device must have the transient power handling capability to last long enough to trip the circuit breaker or primary fuse in the power supply. Removing the primary source of power from the supply ensures that the faulty or improperly adjusted supply is taken off-line and that it will remain off until corrective action is taken.

7-4   Decoupling Capacitors

Decoupling capacitors, also called bypass capacitors, must be used in liberal qualities when advanced TTL or advanced CMOS devices are used.

7-4-1   Local decoupling

Local decoupling capacitors are used to supply local transient-current needs during the period when devices are switching states. Local decoupling capacitors compensate for power distribution system inductance and prevent large transient shifts in local power or reference levels. Even when low-impedance power and ground planes are used, they may not provide sufficient local storage of energy (charge) to supply the instantaneous current (charge) demanded when a large number of devices in a given area switch at the same time. Planes, just as discrete wires, require a finite time for current to flow from one point to another. When several devices in a given area switch at once, significant amounts of instantaneous current must be available to prevent large shifts in local power or ground levels. Even with a well-designed power and ground distribution system, unless local decoupling is present, local transient excursions may exceed specified operating levels. With low supply voltage, devices slow down, signal transitions slow down, and potential system operating speed is reduced. Decoupling capacitors near the power terminals of devices limit local power level transient excursions to harmless levels.

Figure 7-9 represents the output stage of a gate or buffer along with the power source and load circuitry that typically surround a logic device. When $S1$ closes (which represents a gate switching from a *low* level to a *high* level), the nearby 0.1-μF decoupling capacitor ($C1$) supports the local supply V_{cc} level and provides the transient current demanded by the 100-pF load ($C2$). Without the nearby decoupling capacitor, the V_{cc} level at the gate would be reduced for some period of time since the power-supply source inductance L_s impedes the instantaneous current that is needed to resupply the charge transferred to the load. A reduction in V_{cc} may upset other gates in the package. For example, other outputs may shift to incorrect levels. In addition, the gate responsible for the change in V_{cc} (i.e., the gate that switched)

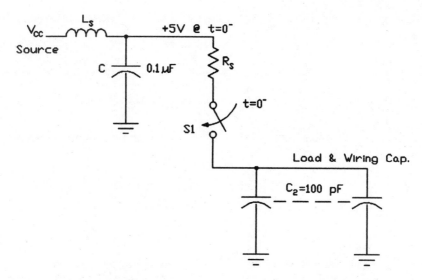

Figure 7-9   Circuit model for a switching device with decoupling.

may not respond as expected because of an out-of-specification V_{cc} level.

Local transient supply voltage disturbances due to local transient load currents can be calculated using the principle of equality of charge. For the circuit shown in Figure 7-9, the final steady-state charge Q_f on $C1$ and $C2$, reached sometime after switch $S1$ is closed, is equal to the initial charge Q_i on $C1$ before $S1$ is closed (see Figure 7-10). That is

$$Q_i = Q_f \qquad (7\text{-}7)$$

$$Q_i = C_1 + C_2(0) \qquad (7\text{-}8)$$

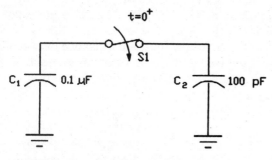

Figure 7-10   Charge transfer from decoupling capacitor to load.

$$Q_f = C_1V_f + C_2V_f \qquad (7\text{-}9)$$

$$C_1V_{cc} = (C_1 + C_2)V_f \qquad (7\text{-}10)$$

$$V_f = \left(\frac{C_1}{C_1 + C_2}\right)(V_{cc}) \qquad (7\text{-}11)$$

For the circuit shown in Figures 7-9 and 7-10, the transient voltage excursion is

$$V_f = \left(\frac{0.1\ \mu F}{0.1\ \mu F + 100\ pF}\right)(5\ V)$$

$$= 4.995\ V$$

$$\Delta V_{cc} = 5\ mV$$

A 5-mV shift will not affect circuit operation. In addition, in an actual circuit, some current will flow through the inductor during the time the device is switching. Thus, actual excursions will be less than 5 mV. However, the above example is a very benign case. In most situations, more than one device in a given package will change at the same time. All eight outputs of an octal driver or register may switch at once, resulting in a transient switching current level eight times that of the above example; such transient-current conditions must be addressed in the design of the system (board) power distribution and decoupling. When high-speed devices are used, it is extremely important to provide an adequate amount of local decoupling for octal drivers or for parts driving large amounts of capacitance.

A general guideline for the amount and the placement of decoupling capacitors for advanced Schottky devices mounted on conventional printed circuit boards, high-speed wire-wrap or welded-wire boards is: One 0.1-μF high-frequency decoupling capacitor mounted as close as possible to the V_{cc} pin of each high-current driver (device) package and each memory device and one decoupling capacitor for every two packages of random logic.

Keep leads as short as possible to minimize the inductance between decoupling capacitors and the device being decoupled.[21] In all applications, the local decoupling capacitance should be at least 100 times the maximum possible simultaneously switched load capacitance C_L. That is,

$$C_d \geq 100C_L$$

Local decoupling capacitors need the following characteristics: excellent high-frequency response, very low equivalent series resistance, and relatively high self-resonant frequency.[22] In most applications, ceramic capacitors are the best choice when size, cost, and frequency response are all considered. Aluminum electrolytic and the various tantalum capacitors do not have adequate frequency response for local decoupling. Where very high frequency response is required, 0.01-μF ceramic capacitors are generally superior to larger value ceramic capacitors.

In summary, local decoupling capacitors supply local transient-current needs and prevent local degradation of V_{cc}. To serve that function, they must be located as close as possible to the power pins of the package being decoupled in order to minimize the interconnection inductance.

7-4-2 Bulk decoupling

Bulk decoupling capacitors help compensate for inadequate power or ground connections between circuit boards and motherboards or between motherboards and power supplies. Large bulk capacitors provide low-frequency replenishment of charge to local decoupling capacitors to help maintain V_{cc} at the proper level.[23]

In addition,

1. Circuit board bulk capacitors reduce the transmission of board-generated noise to the motherboard.
2. Motherboard bulk capacitors help eliminate low-frequency ripple and ringing due to power-supply conductor inductance.
3. Circuit board and motherboard bulk capacitors help reduce the transmission of digital-switching noise back to the power source; thus bulk capacitors help in meeting emission requirements.

In most bulk-decoupling applications, solid-tantalum capacitors that are 10-μF or greater are used. Bulk-decoupling capacitors must be sized for worst-case low-frequency charge replenishment.[24] A good rule of thumb is as follows: The capacitance should be approximately 50 to 100 times the total worst-case simultaneous switched load capacitance, which is the total signal track capacitance plus the total device load for the worst-case combination of signals within a unit that could switch at one time.

A note of caution. Tantalum capacitors, or other large value capacitors, tend to be less reliable than most other devices used in digital systems. Hence, the use of tantalum capacitors should be minimized. It is much more preferable to reduce the inductance of the power distribution system, rather than to try to compensate for a poor distribution system with bulk capacitors. If the impedance of the power distribution system can be kept low enough, high-value (0.1-µF or larger) ceramic capacitors, which are more rugged and have a superior high-frequency response, can suffice for bulk decoupling as well as local decoupling.

In cases where bulk tantalum capacitors are required, careful attention must be given to the application so as to limit ripple current, transient voltage spikes, in-rush current, and polarity reversal.

Tantalum capacitors can be destroyed or degraded unless precautions are taken to ensure that voltage ratings are not exceeded. Tantalum capacitors must be kept as cool as possible since voltage ratings decrease at high temperature. When tantalum capacitors are used for decoupling, they should be selected so that the maximum applied voltage during normal operation is close to the derated maximum rated voltage. Tantalum capacitors should not be excessively derated. However, worst-case operating voltage (including transient conditions) must never exceed the maximum voltage rating at the maximum operating temperature (tantalum capacitors are rated for less voltage at high temperature).

Tantalum capacitors are polarized, so be careful not to reverse the leads during installation. Tantalum capacitors, as well as the location where they are to be installed, should be clearly marked to reduce the opportunity for incorrect installation. Small power-supply reversals at turn ON or turn OFF have been known to damage certain types of tantalum capacitors. Extra care must be taken to ensure that power connections to the unit or board are not reversed.[25] Such capacitors should not be used, or strenuous measures should be taken to ensure that transient power-supply reversals cannot occur. However, reverse-voltage limits cannot be easily met since many of the capacitor styles available in the size range needed for bulk decoupling are only rated for 0.5 V of reverse voltage, and it is difficult to ensure that power-supply undershoots will not exceed that voltage. The use of a normally reversed-biased silicon diode in parallel with a bulk capacitor is not a solution since high-current silicon diodes clamp at 0.8 to 1.2 V, depending upon the temperature and the current. However, there are large value mil style tantalum-cased tantalum capacitors available with reverse-voltage tolerances of up to 3 V, but they are expensive.

Meeting surge-current limits is usually less of a problem than meeting reverse voltage requirements. Slow turn ON of supplies, plus power conductor impedance, is often sufficient to limit surge currents to acceptable levels.

The application of large bulk capacitors presents mechanical as well as electrical problems. The coefficients of expansion of the case, lead frame, and tantalum slug of solid tantalum capacitors are all radically

different. Thus, wide operating temperature ranges and soldering during installation can degrade devices.

For all decoupling needs, for maximum reliability, it is recommended that the largest value capacitor for a given case size and voltage never be selected. Compromises may have been made to pack a
little more into a little less.

7-5 Power Connection Summary

When applying high-speed logic devices, designers must undertake a
thorough analysis of the power distribution system and design the
power and ground interconnections so that switching transients will
not upset the system. The inductance of "real world" power connections cannot be ignored when a system is built with advanced
Schottky or advanced CMOS devices.

7-6 Summary of Techniques for Minimizing
Transient-Current Effects in Power
Distribution Systems

Techniques for minimizing transient-current effects and optimizing
power distribution systems include:

1. Planes should be used to distribute power and ground on circuit
   boards and motherboards. Ground reference planes are always required. Power may be distributed with a ladder or grid configuration but never with point-to-point wiring.

2. A continuous ground plane with no avoidance areas in the IC package or connector pin fields on circuit boards or motherboard is always required.

3. Decoupling capacitors must be located as close as possible to active
   circuitry (i.e., component packages). A good rule of thumb is one
   adjacent decoupling capacitor per high current driver or memory
   device and one for every two packs of random logic (see Section 7-
   4-1).

4. Power supplies should be located as close as possible to the powered
   circuitry, and twisted-pair lines or closely coupled bus bars (to minimize the area between the conductors, and hence the inductance)
   should be used to connect the power source to the local power distribution systems. When power must be remotely located, bulk-
   decoupling capacitors should be used at the power entry points (see
   Section 7-4-2).

5. Ground pins must be evenly distributed across all connector pin

fields (including custom device packages) to prevent local ground upset due to transient currents. The number of connector ground pins required must be determined from a system error budget. A rule of thumb for the minimum number of grounds in a connector is as follows: a minimum of one ground pin per inch of connector length or one per eight output signal lines.

6. When using prototyping boards (such as wire wrap, stitch weld, etc.), the power and ground planes must be continuous through the package- and connector-pin fields and the ground and power connections to component package pins, and board connectors must be made directly to the planes (solder washers or clips must be used on universal boards). Connecting power or ground with pc traces or with discrete wires is not permissible because such connections cause excessive inductance.

References

1. DiCerto, Joseph: "Poor Packaging Produces Problems," *The Electronic Engineer*, September 1970, pp. 91–93.
2. Visco, Anthony P.: "Coaxing Top Bipolar Speeds from Prototyping Boards," *Electronic Products*, September 1, 1987.
3. *FAST Applications Handbook 1987*, National Semiconductor Corp., South Portland, Me, 1988.
4. Blood, William R., Jr.: *MECL System Design Handbook*, 4th ed., Motorola Semiconductor Products Inc., Phoenix, Az., 1988.
5. Saenz, R. G., and E. M. Fulcher: "An Approach to Logic Circuit Noise Problems in Computer Design," *Computer Design*, April 1969, pp. 84–91.
6. *F100K ECL User's Handbook*, Fairchild Camera and Instruments Corp., Puyallup, Wash., 1985.
7. Oates, Edward R.: "Good Grounding and Shielding Practices," *Electronic Design*, No. 1, January 4, 1977, pp. 110–112.
8. Morrison, Ralph: *Grounding and Shielding Techniques in Instrumentation*, Wiley, New York, 1967.
9. McPartland, J. F., and W. J. Novak: *Electrical Equipment Manual*, 2d ed., McGraw-Hill, New York, 1960.
10. *National Electric Code 1981*, National Fire Protection Association, Boston, Mass., 1980.
11. Harper, Charles A.: *Handbook of Electronic Packaging*, McGraw-Hill, New York, 1969.
12. Varney, Douglas: "Determine Wire Size with Nomograms," *Electronic Design*, No. 20, September 27, 1977, pp. 90–93.
13. Harper, Charles A.: *Handbook of Wiring, Cabling and Interconnections for Electronics*, McGraw-Hill, New York, 1972.
14. Lin, H. C.: *Integrated Electronics*, Holden-Day, San Francisco, Calif., 1967.
15. Ormond, Tom: "Backplanes Play a Crucial Role in High-Speed Systems," *EDN*, July 10, 1986, pp. 222–228.
16. *Aluminum Building Wire Installation Manual*, The Aluminum Association, New York, 19XX.
17. *Bimetallic Electrical Transition Joints*, Aluminum Company of America, Pittsburgh, Pennsylvania, 19XX.
18. Harper, Charles A.: *Handbook of Materials and Processes for Electronics*, McGraw-Hill, New York, 1970.

19. Klein, Richard: "Protecting Circuits from Over and Under Voltages," *The Electronic Engineer,* March 1969, pp. 59–61.
20. Fox, Richard W.: "Six Ways to Control Transients," *Electronic Design,* No. 11, May 24, 1974, pp. 52–57.
21. Cowdell, Robert B.: "Bypass and Feedthrough Filters," *Electronic Design,* No. 17, August 16, 1975, pp. 62–67.
22. Harper, Charles A.: *Handbook of Components for Electronics,* McGraw-Hill, New York, 1977.
23. Martin, Arch G., and R. Kenneth Keenan: "Neater Decoupling on Surface-Mount Boards," *Electronic Products,* August 15, 1987, pp. 47–49.
24. Doubrava, Laudie: "Bypass Supply Loads with Care for Optimum Transient Response," *EDN,* September 20, 1979, pp. 113–117.
25. Upham, Arthur F.: "Failure Analyses and Testing Yield Reliable Products," *EDN,* August 8, 1985, pp. 165–174.

Synchronous Design

Synchronous logic design requires that clock inputs of storage elements, such as flip-flops, registers, counters, state machines, etc., only be driven by the system clock.[1,2] Signals generated in combinational logic paths are never used to clock storage elements. Asynchronous inputs such as presets or resets are never used to perform operational system logic functions. Asynchronous inputs are only used for system initialization at power turn ON or for test initialization. Thus, when a synchronous system is in an operational mode, storage elements are only allowed to change states as a result of a clock transition on the clock input and the logic conditions on the control inputs. Asynchronous logic design, in contrast to synchronous design, allows logic signals as well as clock signals to initiate and control the state changes of storage elements.[1,2]

Since it is impractical, if not impossible, when using high-speed logic devices to keep noise levels due to reflections, ringing, crosstalk, and ground bounce (or power-supply droop) to levels that do not exceed TTL or CMOS noise margins, means must be found to tolerate high noise levels.[3] The time quantization and filtering effects of synchronous designs provide that means. Synchronous design maximizes the noise immunity of a logic system by utilizing the clock as a filter. In synchronous designs, devices can only change states following clock transitions. Thus, noise on control signals, except at the critical time near a clock transition, will not cause system upset. Since output transitions are the major source of noise, if the clock period is sufficiently long for the noise to subside before the next clock transition, a clean unambiguous logic decision can always be made.[4] Thus, ideally a synchronous system employs only one clock frequency of a sufficient time period (between active clock edges) to allow for the worst-case noise to settle before the following clock transition. Under such conditions all transients have a chance to subside and quiet signals are available to be sampled on the next clock transition (see Figure 8-1). If there is time for the noise to subside, synchronous systems can tolerate a great deal of switching noise without upset.

Synchronous systems are easy to understand, analyze, test, and change if necessary. In synchronous systems, signals flow in an orderly manner from one clocked element to the next clocked element. Thus, systems are partitioned into manageable and understandable sections. Furthermore, since all clocked elements are clocked with the same clock, a fixed and known quantum of time is established for all operations, which simplifies timing analysis. Test and design verification is also simplified since clock rates can be decreased or increased to determine timing margins without logic malfunctions occurring as a result of special timing paths or requirements.

There are some disadvantages to synchronous designs. System inputs often need to be synchronized to the internal system clock, and the process of synchronization introduces the possibility of creating metastable conditions at the synchronizer (the metastable problem is described in Section 8-1-3). Another perceived disadvantage is that decisions can only be made at clock times, and thus timing paths cannot be optimized, i.e., synchronous designs are perceived to be slower than asynchronous designs.[2] In synchronous designs, the clock frequency must be fast enough to make the fastest decision necessary and all other signal paths are forced to accommodate to that clock time, whereas in asynchronous designs each signal path can be optimized.[1] Thus, asynchronous designs are often promoted for high-speed applications since in some quarter they are perceived as being faster than synchronous designs. Perhaps such an optimization can be achieved in

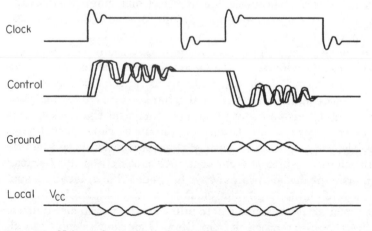

Figure 8-1 Noise and quiet times in a synchronous system.

a benign laboratory environment. However, for production systems that do not have the luxury of using selected parts and that must operate in vastly different environmental conditions over a long lifetime where parts may change characteristics greatly, there is little or no practical speed advantage to asynchronous designs. For a reliable design to be achieved with either a synchronous or asynchronous design, allowances must be made for the worst-case operating speeds of all of the logic components used in the system. When worst-case component timing due to aging, process, and environmental conditions are properly allowed for, there is usually little if any speed advantage to an asynchronous design.

8-1 Design Considerations for Synchronous Systems

8-1-1 Clock requirements

The most crucial part of a digital design is the design of the clock distribution systems. Clock signals cannot tolerate noise or other disturbances in either synchronous or asynchronous designs. Clock signals must be as near perfect as possible (clock distribution is described in Chapter 9). Ideally, there should be only one clock frequency and one phase so that all storage elements are clocked at the same time. Such an arrangement establishes when a system will be noisy and when it will be quiet. If the clock period is sufficient to allow for the worst-case signal propagation path, the system signals will be quiet at the next clock edge and unambiguous decisions can always be made (see Figure 8-1). If multiple clocks, or clock phases, are used, noise will be present more of the time, and for a given system cycle time, the time for noise to subside is reduced.

To minimize the susceptibility of clocked elements with Schottky TTL input levels to noise on clock lines, use only those storage elements that clock on *low*-to-*high* transitions. Clocking on *low*-to-*high* transitions provides more tolerance for coupling or other corruption of clock signals at the peak noisy times immediately following clock transitions since all advanced Schottky TTL devices have a good deal more noise margin in the *high* state (500 mV) than in the *low* state (300 mV). In addition, Schottky devices have more *high*-state noise margin than is specified when they are operated at elevated temperatures, which is where most digital systems operate most of the time. Hence, in high-temperature applications, *low*-to-*high* transition clocking provides additional margin beyond that apparent in the device specifications. *High*-level TTL noise margin is minimum, or nearest the specified value, at the lowest specified device operating temperature (see Chapter 4).

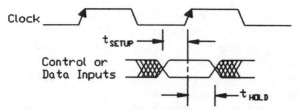

Figure 8-2 Signal setup and hold times.

In either TTL or CMOS systems, all clocked devices should clock on the same edge so that all parts of a system have the maximum possible amount of time to reach a quiet state before the next active clock edge.

8-1-2 Hold-time requirements

Hold time is the time that an input control or data signal to a clocked element must remain stable after an active clock edge (see Figure 8-2). Hold time is the most critical timing parameter in synchronous systems. Hold-time violations most often occur:

1. Where signals go directly from one clocked device to another
2. Where control signals are gated with a clock to form a gated clock
3. Where signals go from fast devices (F, AS, etc.) to slow devices (LS, HC, etc.)
4. Where clock signal alignment is poor

Good clock alignment is essential for preventing hold-time violations. However, perfect alignment can never be achieved. Thus, all signal paths must have some allowance for slight misalignment. Adding extra delay elements, such as additional buffers, or requiring that there be some combinational logic between clocked elements can help alleviate hold-time problems. However, care must be taken since minimum propagation times are generally not specified for the older logic families and are seldom available for the newer logic families for the exact load conditions of a given situation. Minimum propagation times, when specified, are specified with a standard load (50 pF for most advanced Schottky devices). Yet, in many cases, the actual load will be much less than the standard load, particularly when devices are used as delay elements. When the load is less than a standard load, the actual minimum propagation delay will be less than the specified value. Package type can also influence minimum propagation delays. The minimum propagation times shown in the data books for the FAST family are specifically for devices in dual-in-line packages. Devices packaged in packages with less lead inductance, such as leadless chip carriers (LCC), will have shorter minimum propagation times than indicated by the data book values.

Of special concern is the situation where a signal originates from a high-speed part, such as a FAST part, and is received by a slow device such as an LS device. With such a signal flow, i.e., from a fast device to a slow device, it is generally difficult to meet hold-time requirements. Several levels of combinational logic between devices may provide adequate hold time, but in all such cases designers must be careful because of the uncertainty in the minimum propagation times. Off-phase clocking, shown in Figure 8-3, which is undesirable in synchronous systems, is often required in such cases.

8-1-3 Synchronizing asynchronous inputs

Signals coming into synchronous systems are generally asynchronous to the internal system clock and therefore must be synchronized. Most systems must interact with external sources that operate independently of the internal system clock. Signals from independent sources (i.e., with separate clock sources) appear random with respect to the clock in the receiving system. Such signals should be physically isolated from the internal synchronous logic (since the system clock will not act as a filter) and synchronized as soon as possible at one place only. It is important that asynchronous inputs be synchronized at only one point. If multiple synchronizers are used, one synchronizer may have significantly different characteristics than another and may detect an input on different clock edges as the internal clock and the ex-

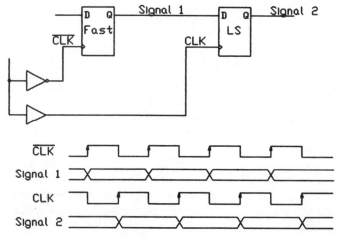

Figure 8-3 Off-phase clocking.

ternal signal shift with respect to one another. Capturing signals at
different times in parallel synchronizers can result in portions of a
system being out of phase with other portions.

At asynchronous interfaces there is always the problem, and the
possibility, that the synchronizing devices will go into a metastable
(unknown or unstable) condition.[5] Metastable operation of digital devices is a malfunction that results in device outputs lingering some-
where in between the two normal output states for some indefinite period. The possibility of metastable operation of clocked devices is
inherent, and such operation is impossible to prevent at asynchronous
interfaces. It is always possible for a bistable device, flip-flop, register,
latch, etc., to go into a metastable state if their input signals do not
meet all of the required specifications for input levels or for setup and
hold times. Since it is impossible to meet such conditions at asynchronous interfaces, make sure that the metastable operation of synchronizing devices does not upset system operation. The recovery time for
a part that goes into a metastable state can be much longer than the
specified propagation delay for the part in a normal operating mode.

In general, faster logic families, such as FAST, have a smaller win-
dow of susceptibility to metastable conditions and recover faster.
Studies have shown that a number of the more common FAST devices
have very good metastable recovery characteristics.[6] The probability
of recovery is very high after 20 ns. Similar studies have shown LS
parts to have very poor recovery characteristics. Thus, FAST devices
are generally recommended for synchronizers at asynchronous interfaces. Programmable devices such as PALs and PLAs should never be
used to synchronize inputs. Their performance under such conditions
is not predictable (they tend to have a wide range of specified susceptibility), and there is a tendency to overlook the possibility of asynchronous signals being routed to multiple clocked devices and synchronized at multiple places.

Where asynchronous buses are received, it is best to capture and
synchronize the strobes or control signals (assuming that the timing of
the bus is such that it is possible to do so) rather than to synchronize
an entire bus. Such an arrangement minimizes the chance for metastable operation, since only the control signals must be synchronized.
The entire bus is not subjected to the possibility of incorrect setup or
hold times. Attempting to synchronize an entire bus widens the win-
dow of susceptibility to metastable operation and increases the possibility that one or more bits will go into metastable operation; all of the
receiving devices will not be identical—some will be slower and some
will be faster than others, and the clock phasing may be slightly different at each device.

8-2   Miscellaneous Considerations for Synchronous Designs

Certain devices inherently do not fit into synchronous logic designs. They include one-shots (multivibrators), transparent latches, and master-slave devices.

8-2-1   One-shots

One-shots should not be used in synchronous systems. One-shot output changes are not clock controlled. Thus, one-shots clearly violate the synchronous design requirement that all signals change relative to the active edge of the system clock. In addition to violating synchronous design requirements, designers have several practical reasons for not using one-shots in modern digital systems. One-shot time-out periods are difficult to control or determine accurately and as a result worst-case time-out limits are generally greatly underestimated. One-shots are also difficult to test. Most automatic testers used for digital hardware testing cannot test one-shots.

However, there is often a temptation to use one-shots when long time delays or long time-out periods are needed and board space is at a premium. Long time-out periods using standard 4-bit counters may require a number of packages to achieve the same time-out period that can be achieved with a single one-shot (package), a resistor, and a capacitor (most one-shots require an RC timing network—shown in Figure 8-4). However, one-shots with long time-out periods often have much greater variations in their time-out periods than expected. One-shots with long time-out periods require RC timing networks with large value resistors and capacitors. Yet, large value resistors and capacitors with close tolerances are generally not available, and they tend to have large tolerance variations with temperature and aging. For example, 20-μF tantalum CLR79 capacitors are available with initial tolerances of ± 5 percent, but CLR79 capacitors can be expected to vary an additional -32 to $+12$ percent with temperature and an

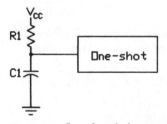

Figure 8-4   One-shot timing network.

additional ±15 percent variation can be expected with stress and aging. Standard carbon resistors are available with ±5 percent initial tolerances, but an additional ±15 percent variation can be expected with aging and solder stress. Thus, timing capacitor variations can be as large as −52 percent (−5, −32, −15 percent) to +32 percent (+5, +12, +15 percent), and resistor variations can be as large as ±20 percent (±5, ±15 percent).

For example, a 200-kΩ carbon timing resistor with a ±20 percent tolerance will vary from

$$R_{MIN} = 160\ \Omega \quad \text{to} \quad R_{MAX} = 240\ k\Omega$$

and a 20-μF CLR79 timing capacitor with a +32, −52 percent tolerance will vary from

$$C_{MIN} = 9.6\ \mu F \quad \text{to} \quad C_{MAX} = 26.4\ \mu F$$

and the two possible extremes in time-out period are

$$T_{MIN} = 160\ k\Omega \times 9.6\ \mu F = 1536\ ms$$

$$T_{MAX} = 240\ k\Omega \times 26.4\ \mu F = 6336\ ms$$

The possible variation in time-out period is

$$\frac{6336\ ms}{1536\ ms} = 4.125$$

A four-to-one variation in the time-out period is more than most applications can tolerate, and there are other sources of time-out error that must also be considered. Internal one-shot circuitry is not perfect; it has uncertainty and variation which are generally not specified. Leakage currents also affect one-shot time-out periods and in some cases can be of such a magnitude that time-out never occurs.

Several sources of leakage current contribute to time-out errors:

1. *Capacitor leakage current:* The worst-case leakage-current specification for CLR79 capacitors is 1.5 μA (some other styles of capacitors have much larger leakage).

2. *Board leakage current:* Board leakage depends on environmental conditions and the physical arrangement of the interconnections. It is not uncommon for board impedance to be as low as 100 kΩ in normal high humidity conditions and as low as 10 kΩ in some extreme salt air conditions.

3. *One-shot timing port input current:* One-shot loading of the timing network is usually not specified (yet there is some loading), and thus its effect on time-out period is unknown. (Note: One-shot input

current is technically not a leakage current, but the effect is the same.)

If all the leakage sources are near maximum and a high value timing resistor is used, time-out may never occur. High leakage currents may load the RC timing network to the extent that the RC network never reaches a level sufficient to trigger the one-shot, and as a result the one-shot never times out (see Figure 8-5).

With the possibility that one-shots with long time-out periods may not time out, or if they do time out, the variation in time-out period may be as large as four-to-one, long time out one-shots should be avoided if at all possible.

When long time-out periods are needed, there are a number of microprocessor support chips that have long counter chains, and there are a number of long ripple counters (in a single package) in the HC/HCT logic family. Counters are a better solution than one-shots in most applications requiring long time-out periods.

8-2-2 Latches

Latches, depending upon how they are used, may violate the requirement that all signals change relative to the active edge of the clock. Latches also may impose latch control signal restrictions that are difficult to meet in a synchronous, single phase, clock system. It is important that latch control signals be timed so that hold-time violations do not occur at latch inputs or at devices receiving latch outputs.

8-2-3 Master-slave devices

Master-slave devices reduce the amount of quiet time in a system. Since master-to-slave transfers are completed on the off-phase edge of the clock, signal settling time is reduced by one-half (for a 50 percent duty cycle clock).

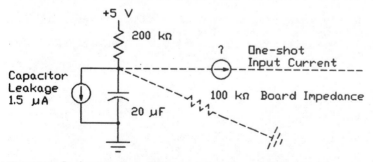

Figure 8-5   Leakage currents that can prevent one-shot time-out.

8-3   Summary of Synchronous Design Practices

In synchronous designs, storage elements can only change states as a result of a clock transition on a clock input.

1. Clocked elements are only clocked by the system clock.

2. Only a single-phase single-frequency clock should be used.

3. All clocked elements should be edge triggered (as opposed to latches or master-slave devices).

4. Clocked devices with TTL input levels should be selected so that all devices clock on the *low*-to-*high* clock transition.

5. Asynchronous presets and resets on clocked elements should never be used for performing operational system logic functions because of their susceptibility to noise.

6. Unclocked feedback paths (such as cross-coupled gates) should not be used because they are susceptible to being upset by system noise.

7. Never use a counter carry or a decoder output as a clock (such outputs are expected to have spikes).

8. Do not route asynchronous signals to multiple points within a functional unit until they are buffered and synchronized to the internal clock.

9. Monostable multivibrators (one-shots) should not be used.

References

1. Mano, Morris M.: *Digital Design*, Prentice-Hall, Englewood Cliffs, N. J., 1984.
2. Fletcher, William I.: *An Engineering Approach to Digital Design*, Prentice-Hall, Englewood Cliffs, N. J., 1980.
3. Funk, Richard, and James Nadolski: "Advanced CMOS—Pinouts Are Not the Crucial Factor," *Electronic Engineering Times*, Monday, August 4, 1986.
4. Hayes, John P.: *Computer Architecture and Organization*, McGraw-Hill, New York, 1978.
5. Kleeman, L., and A. Cantoni: "Metastable Behavior in Digital Systems," *IEEE Design & Test of Computers*, December 1987, pp. 4–19.
6. Chaney, T. J.: "Measured Flip-Flop Responses to Marginal Triggering," *IEEE Transactions on Computers*, Vol. C-32, No. 12, December 1983.

9

Clock Distribution

High-quality clock signals are essential for high-speed high-performance digital systems. Synchronous systems do not require precise control of data and control signals, but clock signals must be as nearly perfect as possible. Clock-signal wave shape and skew must be controlled, and a high level of noise immunity must be assured. Thus, clock-signal distribution must receive a great deal of attention. Well-conceived and -executed clock distribution schemes provide solid foundations for reliable high-quality systems. On the other hand, poor clock distribution schemes result in systems that require a great deal of patching and at best operate with little margin.

9-1 General Considerations

To ensure system synchronization and optimize noise tolerance, equip systems with only one clock source and use only one clock frequency and phase (for the reasons described in Chapter 8). In practice, a single clock frequency and phase is often not possible for numerous reasons, e.g., speed limitations of certain necessary components or interfaces. However, the goal should be to minimize the number of clock frequencies and phases. Special attention must be given to areas of systems that operate with multiple clock frequencies or that communicate with other components or other areas that operate with different clock frequencies.

Clock signals require an optimal electrical environment. Clock-signal quality, uniformity of propagation delays (skew), and noise tolerance are issues that must be addressed throughout clock distribution networks. Clock signals must be as nearly perfect as possible. Excessive ringing and glitches cannot be tolerated on clock lines. Thus, clock distribution circuitry must incorporate some form of line termination for wave-shape control.[1] Clock signals must be routed to their destinations using reasonably well controlled impedance paths so that proper termination can be selected.

Clock distribution networks (trees) must be designed to minimize skew between clock signals. To accomplish the needed alignment, standardize clock distribution circuitry. Propagation delays must be matched throughout clock distribution networks. Clock drivers at each level must be identical generic devices of the same logic family, and where possible, all drivers in a given level should be located in a common package. Drivers and loads must be arranged so that loads are equalized. Clock lines must be very short so that wiring propagation delays are insignificant, or lines must be of near equal length in each distribution level to equalize physical propagation delays.

Clock lines must be physically arranged so as to minimize cross coupling or other adverse interference. Parallel runs adjacent to other signals or other clocks must be kept as short as possible to minimize the opportunities for cross coupling. When multilayer printed circuit boards are used, clock signals should be isolated and shielded from other signals by locating the clock traces between reference planes. On welded-wire or wire-wrap circuit boards or backpanels, each clock line should be twisted with a ground line to provide some shielding and control of line impedance. Make sure that clocks lines do not get bundled with other signals.

When advanced Schottky or advanced CMOS devices are used as clock drivers, dc drive limits are of little significance. When advanced Schottky or advanced CMOS devices are used as clock drivers, they should never be expected to drive more than a small fraction of the rated dc load. Individual driver and total package load must be limited to minimize switching currents in package power and ground pins which will in turn minimize ground bounce or power-supply droop. As a rule of thumb, the number of loads driven by any one driver should be limited to six and the total package load limited to thirty-six. If octal parts, such as F240 or F244s are used, no more than six of the eight drivers (those nearest to the ground pin) should be used. The two spare drivers should not be used for other purposes. Even with limitations on the loading of the active drivers, six drivers switching six loads at once can generate significant ground bounce (or power-supply droop), which will appear as spikes on the outputs of the other drivers in the package.

Different clock frequencies or phases should not be mixed in a common package. If two different frequencies are mixed in a common package, spikes due to ground bounce or power-supply droop may appear on the lower frequency when the higher frequency switches (see Figure 9-1).[2]

To supply the transient current needs of clock drivers, local decoupling capacitors and circuit boards with continuous low-

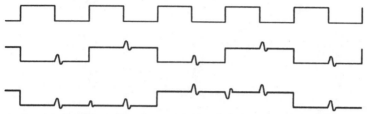

Figure 9-1 Possible spikes due to mixing clock frequencies in a common package.

impedance power and ground planes are a necessity.[3] Where prototyping boards, such as stitch weld or wire wrap, are used, "Schottky" boards with continuous (no cut-out areas around pin fields) low-impedance power and ground planes must be used.[4] Clock-driver package and decoupling capacitor power and ground connections must be made directly to the circuit board power and ground planes. Where prototyping boards, such as stitch weld or wire wrap, are used, the connections must be made with solder clips or washers to minimize the inductance of the connection.[5] Actual welded-wire or wire-wrap wiring should never be used to make clock driver power or ground connections. Each clock-driver package should be decoupled with a high-frequency capacitor (typically 0.1 μF) located as close as possible to the package (see Section 7-4-1 for guidelines on selecting the value).

To minimize the effects of signal degradation or cross coupling into local clock signals, select only clocked devices that change states on the *low*-to-*high* transition of the clock (see Section 8-1). The goal is to have all local clock signals in the *high* state, which has maximum noise immunity, when the majority of system signals are switching and generating noise.

When devices with TTL input levels are used for clock distribution, clock-signal phasing at all major distribution levels in clock trees (i.e., at all major interfaces) should be such that clock signals switch from the *low*-to-*high* level at the same time that component-level clock signals transition from *low*-to-*high* level. The design objective is to ensure that long clock lines that have a high probability of being exposed to noise, such as the clock lines that traverse backpanels or motherboards, are in the maximum noise-immunity state during the time of peak system noise (Figure 9-2).

When TTL systems use low-frequency clocks (10 MHz or less), consider shaping clock signals so that the low state is as short as possible to optimize clock-signal noise immunity (see Figure 9-2). Such an arrangement keeps clock signals in the maximum noise margin state for a longer period, allowing a longer time for system signals to transition and a longer time for noise due to delayed transitions to die out before

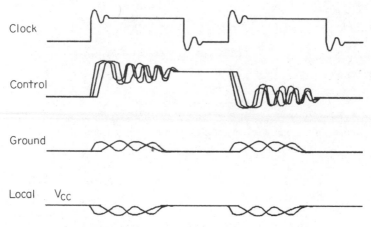

Figure 9-2   Clock phasing for maximum TTL-level noise immunity.

clock signals go to the *low* state, which has less noise margin than the *high* state. Above 10 MHz a nonsymmetrical clock signal is usually not practical, but for systems with slower clock rates such a configuration should be considered. If clock pulses are narrowed, care must be taken to ensure that clock pulse width is adequate for the slowest logic family or device that might be used in the system.

When clock sources and loads are not solidly referenced to a common uninterrupted ground plane, clock signals must be transmitted by some technique that provides a high degree of noise margin. The noise margin of standard single-ended TTL or CMOS devices is not adequate for clock distribution when the transmitting and receiving devices are not directly referenced to a low-impedance common ground plane. When an adequate low-noise reference system does not exist, clock signals are typically distributed using some form of balanced differential transmission that has a high level of common-mode noise rejection capability (see Section 9-2-1).

Test clock input port.   Each major system unit should have a test clock input port to allow units to be tested for clock operating frequency margin. It is important to test for operating margin during design verification and during production. A test port also allows systems to be operated at reduced clock rate during initial check out so that logic errors can be corrected before the additional complications of full-speed operation are encountered. A recommended circuit for test clock ports is shown in Figure 9-3. Resistor $R1$ and diode $D1$ are to protect the gate in case a high drive pulse generator is used for the clock source and the clock source is turned ON when gate V_{cc} is not present. As shown the test port is activated when the TEST pin is jumpered to ground.

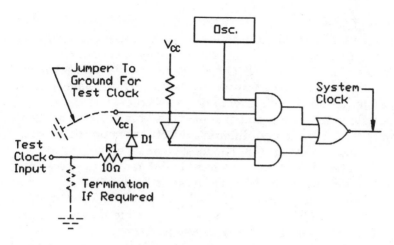

Figure 9-3   Test clock input port circuit.

9-2   Clock Distribution Categories

There are three distinct categories of clock distribution in the typical large digital system:

1. Unit-to-unit connections (i.e., between chassis, racks, cabinets, etc.)
2. Board-to-board connections (i.e., motherboard clock distribution)
3. Component-to-component connections (i.e., circuit board clock distribution)

Interconnection categories 2 and 3 are present in most systems. The need for category 1 depends upon system size and the functional relationship of separate units, if there are any. Small self-contained systems have no need for unit-to-unit connections. Likewise, many large systems are composed of separate autonomous functional units, each separate unit having its own internal clock source.

9-2-1   Unit-to-unit clock connections

High-frequency clock connections between physically separated units (that is, between units that do not have a common backplane, backpanel, or motherboard) is one of the most difficult clock distribution tasks. When clocks must be sent between components that are not directly referenced to a common ground plane, they must be transmitted by some means that provide a sufficient amount of noise tolerance for the application. In Chapter 6, the high-frequency transient switching currents of advanced Schottky devices are shown to result in the generation of significant voltage transients across the inductance of a

1-in length of wire. It follows that the inductance of several feet of wire, as is generally required to interconnect separate units, will result in very large transient voltage shifts that exceed the noise margin of single-ended TTL (or CMOS) devices. Thus to prevent noise from corrupting clock signals transmitted between units, some form of transmission that has a high level of noise rejection is needed.

For most high-frequency unit-to-unit clock transmission applications, some form of balanced differential network must be used to meet the needed noise tolerance. The most common arrangement consists of differential line drivers connected to differential line receivers with twisted-pair interconnections as shown in Figure 9-4.

For unit-to-unit applications with clock rates up to 8 MHz, 26LS31 line drivers and 26LS32 line receivers are possible device choices. Above 8 MHz, 26F31 or 26C31 drivers and 26F32 or 26C32 receivers are required; 26LS parts will overheat and fail if operated much above 8 MHz. The 26F31 and 26F32 are suitable for applications requiring clock rates of up to 25 MHz, and the 26C31 and 26C32 to 40 MHz (depending on operating temperature—the C parts are only available in the −40 to +85°C range at present).

The various 26 XX 32 receivers have the ability to function correctly in the presence of ±7 V of common-mode noise or dc offset, which is adequate for most applications. For very noisy applications, a modified version of the basic 26 XX 32, the 26 XX 33, with ±15 V of common-mode rejection is available. The 26 XX 33 has the same basic internal circuitry as the 26 XX 32, but the 26 XX 33 has a larger ratio input divider network which allows operation over a wider input voltage range. The penalty is less sensitivity (i.e., the minimum required difference in the two input levels for a defined output), 0.5 V versus 0.2 V for the 26 XX 32, but in many applications a 0.5-V input sensitivity is more than adequate.

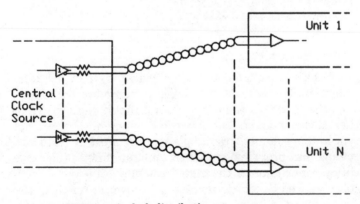

Figure 9-4   Unit-to-unit clock distribution.

Where clock frequencies greater than 40 MHz must be transmitted between units, custom circuits or adaptations of high-speed analog integrated circuits are required. Off-the-shelf differential line drivers or receivers for ≥ 40 MHz are not available. Most high-speed differential line-driving applications require custom driver designs. The situation is somewhat better when high-speed receivers are needed; analog comparators, such as the AM686 or AD9686 (see Section 10-3-2 and Figure 10-6), can serve as high-speed digital receivers. However, in most receiver applications, input divider networks are needed to enhance analog comparator common-mode noise rejection capability. The common-mode rejection range of the typical analog comparator is no more than ± 3 V. Yet, most unit-to-unit communication should be able to tolerate ± 7 V or more of common-mode voltage depending upon the application. Input divider networks can be used to increase common-mode rejection, to terminate lines, and to provide some transient or fault condition current-limiting protection for comparator inputs. Clamp diodes can be used in conjunction with input divider networks to provide additional transient or fault overvoltage input protection.

Since most unit-to-unit interconnections are long transmission lines (see Section 11-5), most unit-to-unit lines must be properly terminated to ensure clock-signal quality. To allow proper termination, transmit unit-to-unit clock signals using dedicated driver-receiver pairs connected with controlled impedance twisted-pair lines. Unit-to-unit clock lines should be terminated so that they are slightly underdamped for optimum edge speed but not so underdamped that excessive ringing occurs. Underdamped signals have faster rise times and thus reach threshold levels faster than critical damped or overdamped signals. A suggested range for termination impedance is 1.2 to 2 times the characteristic impedance Z_o of the line being terminated. Series termination is recommended for most unit-to-unit clock connections. Series termination provides short-circuit protection for drivers, dissipates less power than load termination, and reduces unit-to-unit current flow.

In certain very noisy applications, fiber optics may provide the most appropriate unit-to-unit clock interconnection.

Summary of unit-to-unit clocking techniques

1. Centralize the source and distribution point for system clock signals.

2. Use balanced differential driver-receiver interconnections for noise immunity (single-ended interconnections are not adequate).

3. Use a separate driver-receiver and shielded twisted-pair line for each unit-to-unit clock.

4. Terminate all lines to ensure clock-signal quality.

5. Use equal line lengths to balance propagation delays.

6. Use the same type of device for all drivers to balance propagation delays.

7. Use the same type of device for all receivers to balance propagation delays.

9-2-2   Board-to-board clock connections

Board-to-board clock-signal interconnections, where the individual circuit boards are referenced to a common backpanel or motherboard ground plane, are generally transmitted and received using single-ended devices. Single-ended interconnections are used for economical and practical reasons. Single-ended drivers are readily available at low cost, and single-ended interconnections use the minimum number of connector pins. However, providing the electrical environment needed for single-ended clock signals requires a great amount of design effort. Single-ended devices do not have a great deal of noise margin, and board-to-board clock lines tend to be long and to run through noisy areas. When single-ended interconnections are used, it is essential that circuit board-to-motherboard connectors have an adequate number of properly distributed ground pins to ensure low-impedance board-to-motherboard ground connections (see Chapter 6). Single-ended board-to-board clock-signal lines must be properly terminated, and they must be routed properly in order to reduce the possibility of coupled noise. Noise on clock lines that run between boards is a common system problem.

Clock signals for boards common to a backpanel or motherboard should all originate from a central source. In the typical application, the central source consists of a clock board that houses a master oscillator and a clock buffer tree to drive clocks to other boards on the common motherboard. In some applications (those with a category 1 level of distribution), instead of housing an oscillator, the central clock board receives an externally generated clock and redistributes it to the local boards. From the central source, an individual clock signal should be sent to each circuit board on the backpanel using a dedicated clock driver and receiver for each board in the unit as shown in Figure 9-5.

Clock signals should never be run to multiple boards when advanced Schottky or advanced CMOS devices are used because of the difficulty in properly terminating multiple board loads. Multiple board load configurations often experience operational problems when one or more boards (loads) are not present or when a board is extended out of the housing for troubleshooting purposes. Either situation

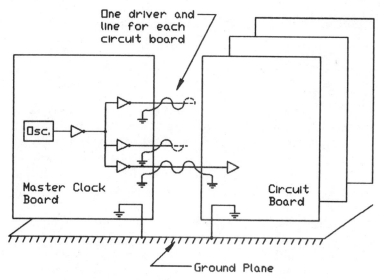

Figure 9-5   Board-to-board clock distribution.

changes the physical arrangement of multiple load clock-signal lines
and will upset the termination of such lines. To achieve the clock-
signal quality needed for proper system operation, properly terminate
board-to-board clock lines at all times when advanced Schottky or advanced CMOS devices are used. Likewise, all systems must be designed so that they will operate when boards are absent or when extender boards are in use. All systems will experience component
failures or other faults and must therefore be able to operate with
some number of circuit boards extended out of the housing so that test
equipment can be connected to board-level signals. The only practical
way to achieve such a design is to send an individual clock line to each
board. Each board should receive its clock signal with a single receiver that can be used to buffer and redrive the clock within the circuit board. Clock lines from the central source should never be used to
directly drive devices on circuit boards; such an arrangement is certain to degrade the clock signal and result in improper system functioning.

Be careful when routing clock-signal lines on backpanels or
motherboards. Clock interconnections must be implemented with controlled impedance lines so that lines can be properly terminated and
so that crosstalk can be controlled. On wire-wrap backpanels, clock
lines should be run twisted with a ground wire as shown in Figure 9-6.
The ground line should be connected to ground at both the source and
the load, as close as possible to the actual driver and receiver, to provide a direct return current path. However, in most applications it is
not practical to run ground lines to the actual drivers and receivers on

Figure 9-6   Twist clock lines with a ground line.

circuit boards. A ground pin on the source and load circuit board–to–motherboard connectors near the active clock line exit on the source board and its entrance on the load board is close enough for most applications. A twisted-pair interconnection provides a known and controlled characteristic impedance (around 100 to 120 Ω) and some shielding from adjacent signals. Make sure that clocks lines do not get bundled with other signals.

Clock-signal interconnections should not have significant discontinuities (in characteristic impedance) or large branches or stubs (and there should not be any if one source and one load are present). Large discontinuities, branches, or stubs can cause severe reflections which may degrade clock-signal edges to the extent that false clocking occurs.

To control crosstalk, physically isolate clock-signal traces on printed circuit motherboards from other signals. If possible, they should be isolated in the vertical direction by reference planes. In the horizontal direction, line-to-line spacing between traces with different frequency clock signals or between clock signals and other signals must be such that little coupling occurs. As a rule of thumb, the line spacing should be such that the coupling does not exceed 50 percent of the worst-case noise margin of the clock receivers (see Section 11-4). A clock-signal noise budget that includes an allowance for crosstalk should be established for each design.[6]

In all cases, board-to-board clock lines should be as short as possible, but where clock lines cannot be kept very short, all lines should be made equal in physical length so that line propagation delays are matched. To minimize the length of board-to-board clock lines, locate the clock source board near the center of the motherboard or card cage. All drivers and receivers should be of the same generic type and logic family, and the terminations and the effective load at each receiving board should be the same to minimize skew in the active clock edge.

When advanced Schottky or advanced CMOS devices are used, clock lines that traverse backpanels must be terminated to ensure clock quality. Such lines generally have two-way propagation delays that are long relative to the rise (or fall) time of the signal and hence must be treated as transmission lines. Either source or destination termination may be used for TTL-level clock signals. Each method has advantages and disadvantages, but destination termination, shown in Figure 9-7, is recommended for most TTL applications. When CMOS devices with true CMOS input threshold levels are used, split termination cannot be used; split termination prevents *high* inputs from reaching proper *high* CMOS levels. Thus, CMOS board-to-board clock signals must be source-terminated as shown in Figure 9-8.

Source termination can be implemented with small-value series resistors or ferrite shield beads located very near the driver. One advantage of series termination is that no additional dc power is dissipated as a result of termination. However, in TTL applications, make sure that *low*-level noise margin is not significantly reduced by series-termination resistors. In CMOS applications, the dc drop across series-terminating resistors is not significant since CMOS inputs source or sink very little current. To minimize TTL noise margin loss and to provide underdamped signals, use series-termination resistors that are approximately one-third to one-half the line characteristic impedance Z_0. A value in the range of 22 to 33 Ω is appropriate for most applications. The value should be low enough so that the *low* state noise margin is not reduced by more than 50 to 100 mV. When determining the response of a source-terminated transmission line, include driver output impedance. The effective source impedance is the

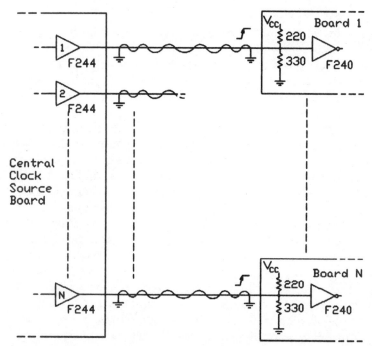

Figure 9-7   Board-to-board TTL-level clock distribution.

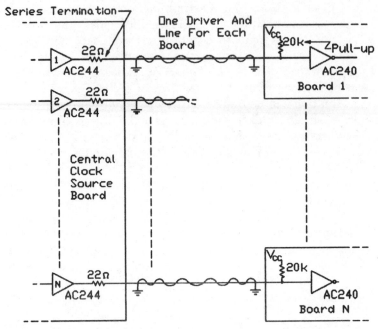

Figure 9-8   Board-to-board CMOS-level clock distribution.

sum of the driver output impedance and the external series-terminating resistor.

Ferrite beads are recommended when TTL clock signals must be series-terminated.[7] Ferrite beads, which have zero static impedance, eliminate possible dc offset problems yet provide the needed amount of impedance at high frequencies.

Destination termination of TTL-level backpanel clock lines is customarily implemented with split terminations (sometimes described as a Thevenin's termination). For optimum response, the Thevenin's equivalent impedance of split-termination networks should be slightly higher than the line characteristic impedance Z_o. For most pc board and welded-wire or wire-wrap applications, split termination networks with resistor values of 220 and 330 Ω, as shown in Figure 9-7, are used.[1] The equivalent impedance of a 220-330 Ω split termination is 132 Ω, which provides a near match for welded-wire or wire-wrap boards and approximately a two-to-one underdamped termination for pc boards (which is acceptable); 220-330 Ω networks are available in standard multiple resistor packages. When the impedance of split terminations are selected, a trade-off must be made between driver current sink requirements and the closeness of the match to line impedance. Normal TTL devices do not have a great deal of current drive capability, but do have high current sink capability. Thus, termination networks must be optimized for minimum loading of driver in the *high* state. Split termination accomplishes that purpose. No high-level drive current is required when split termination resistor values are properly selected. The divider effect of the split termination network establishes the *high* level. In the *low* state the driver must sink a current equal to V_{cc} divided by the value of the termination network resistor connected to V_{cc} (220 Ω in the case of a 220-330 Ω network). When 220-330 Ω networks are used, the sink current is about 22 mA under nominal conditions. Since the standard FAST or AS driver is only rated for 20 mA, drivers with high-current capability, such as F240 or F244, must be used to meet the current sink requirements of a 220-330 Ω split termination. The advantages of destination termination include:

1. Coupled noise and crosstalk tend to be damped more quickly.

2. Edge rates do not degrade as much as they would under similar circumstances (line and load impedance) with source termination.

3. More than one device can be used to receive the signal (on the receiving board), which helps minimize board clock tree depth on large boards.

The disadvantages of destination termination include:

1. Increased power dissipation

2. Higher current drive requirements placed on the driver

In most applications, the advantages outweigh the disadvantages. Thus, destination termination is recommended for most board-to-board applications with TTL-level interfaces. Destination termination cannot be used on CMOS interfaces.

With either source or destination termination, clock lines should be terminated so that they are slightly underdamped for optimum edge speed but not so underdamped that excessive ringing occurs. Underdamped signals have faster rise times and thus reach threshold levels faster than critical damped or overdamped signals. A suggested range for termination impedance is 1.2 to 2 times the characteristic impedance Z_o of the line being terminated.

The phasing of TTL-level clock signals that traverse backpanels or motherboards should be such that the *low*-to-*high* transition of the clock signal is the system's active clock edge. Such a phasing will ensure that the clock lines are in the *high* state with maximum noise margin when the majority of system signals are switching. Systems with clock rates that are relatively slow with respect to the minimum clock pulse width requirements of the devices being clocked should shape clock signals so that the logic *low* state is as short as possible to further enhance noise margin. Such an arrangement keeps clock signals in the maximum noise margin state for a longer time; thus, a longer time is allowed for system signals to transition and a longer time for noise due to delayed transitions to die out before clock signals go to the *low* state, which has less noise margin than the *high* state.

The need for a large number of well-distributed grounds in board-to-motherboard connectors is often overlooked. It is important to solidly reference the clock distribution board to the motherboard ground plane so as to ensure that the clock distribution board does not experience excessive ground bounce when all clock drivers switch simultaneously. An admirable design goal is to provide an adjacent board-to-motherboard ground pin for each clock-signal pin in the connector. In most cases that is not practical; connector pin limitations often force a compromise in the number of ground pins. If it is not possible to provide a ground pin for each clock signal, perform a careful analysis to determine the number of ground pins required to keep board ground bounce within limits. The design task is to keep clock source board transient reference-level shifts due to ground bounce below the threshold level of the most sensitive clock receiver or receiving device on the clock source board.

When board-to-motherboard connectors do not contain an adequate number of ground pins or when other noisy reference conditions exist, some form of balanced differential connection to the clock source (if it is not on the central clock board) and loads must be used to ensure clean, noise-free clock transmission. If standard TTL or CMOS interfaces are used when board-to-motherboard connectors do not contain an adequate number of ground pins, noise may be injected into the clock signals, and noise cannot be tolerated on clock signals. Thus, if single-ended clock inputs are to be used, a low-impedance reference path must exist between board reference planes and the motherboard reference plane to prevent reference-level shifts in excess of single-ended TTL or CMOS noise margins.

Summary of board-to-board clocking techniques

1. Centralize the source and distribution point for board-to-board clock signals.

2. Use a separate clock driver-receiver and clock line for each board.

3. Terminate all clock lines to ensure clock-signal quality.

4. Use very short lines or make all lines equal in length to keep line propagation delays equal.

5. Twist each clock line with a ground line to reduce coupling when wired backpanels are used.

6. To optimize noise margin, phase TTL-level board-to-board clock signals so that *low*-to-*high* transitions are in phase with the active edge of the clock at the component level.

7. Use the same type device for all drivers to match propagation delays.

8. Use the same type device for all receivers to match propagation delays.

9. Limit the number of drivers or receivers used in octal packages to keep ground bounce within limits.

10. To prevent system upset due to ground bounce, do not mix clock frequencies, phases, or other signals in clock driver or receiver packages.

11. Provide a decoupling capacitor for each clock driver and receiver.

12. Connect power and ground pins of clock drivers and receivers directly to the circuit board power and ground planes (no wiring).

9-2-3   Board-level clock distribution

Each circuit board should be designed to operate with only a single clock frequency and phase. A single frequency and phase is not always possible, but that should be the goal. Each board clock input signal (or signals if there are more than one) should be received with a single receiver located near the input connector of the board and then fanned out to the loads as shown in Figure 9-9. Small systems with only one

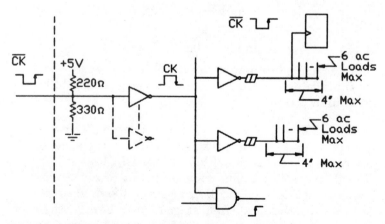

Figure 9-9   Circuit board clock tree.

circuit board may have an on-board oscillator, but regardless of whether an on-board clock source or an external one is used, a standard clock tree, such as the one shown in Figure 9-9, should be used to buffer and fan-out clock signals on the board or smallest self-contained unit. Incoming clock lines on a circuit board should be terminated as close as possible to the load (receiver) if load termination is used; the receiver and termination should be located as close as possible to the point where the clock signal enters the board (see previous discussion on board-to-board clock distribution, Section 9-2-2).

A board-level clock tree (shown in Figure 9-9) provides a means of fanning out to multiple loads on a board without presenting a large load to the clock source. It is desirable to use only one device to receive input clock signals so as to minimize the capacitive load at the end of long backpanel lines. However, the requirement that board clock inputs be limited to one load may force board clock trees to extra levels on large boards, thereby creating large fan-out needs. Since extra levels add uncertainty to clock edge phasing, they are undesirable in clock trees. Thus, in applications where boards are very large, adding more receiving devices is usually a better solution than increasing clock tree depth. Up to six receivers can be used with 220-330 Ω split terminations, but the number should be kept as small as possible. When source termination is used, only one receiver should be used.

Alignment of board clocks is critical in synchronous systems. Active clock edges must be closely aligned to prevent hold time violations when register-to-register transfers must be made. To optimize alignment of board clock signals, take the following steps.

1. The final level of the board clock tree buffers should be centrally located.
2. Identical generic devices, in the same package where possible, should be used for buffers in each given level.
3. Loading for each level should be balanced.
4. Line lengths should be similar in each level. For very large boards, definite minimum and maximum clock line lengths should be established to limit possible skew.

Ground bounce can be a severe problem in clock buffer packages since all drivers in clock buffer packages switch at the same time. When all devices in a package switch simultaneously, load and totem-pole feedthrough currents combine and can result in severe ground bounce. Ground bounce is of particular concern in the final stage of a clock tree since each driver in the final stage tends to be driving heavy capacitive loads. To minimize the potential for ground bounce disturbances, the number of clock buffer loads should be limited to well below dc fan-out ratings. For example, standard FAST devices have a dc drive rating of 33 standard FAST inputs. However, 33 loads at 5 pF per load results in a 165-pF load (with no allowance for the wiring capacitance), which is well above the 50-pF load at which the ac device parameters are specified. In addition, if all eight devices in an octal driver package, such as a F240 or F244, were to be used, with each buffer driving 33 loads, the total ac load that would have to be charged or discharged at each clock-edge transition would be 1320 pF (165 pF × 8). Such a large ac load would result in severe ground bounce. Ground bounce, when an octal buffer package is used as a clock driver, is further aggravated by the totem-pole transient feedthrough current of eight simultaneously switching buffers (see Chapter 3). The use of eight fully loaded buffers is somewhat an extreme example, but the ac load on clock buffers and clock buffer packages must be limited. A good rule of thumb when using advanced Schottky or advanced CMOS devices is to limit the ac load to no more than six loads per driver and no more than 36 loads per package. When octal devices with a single power and ground pin are used as clock buffers, it is best to use no more than six of the drivers in the package and the six used should be the six nearest the ground pin. The extra drivers in octal clock buffer packages should not be used for other purposes since they may exhibit severe ground-bounce spikes during clock transitions. It is important that they not be used for signals that might go to asynchronous inputs, such as flip-flop sets or resets, that are sensitive to spikes. In cases where multiple clock frequencies or phases must be used on a board, they should not be mixed in common buffer packages. Ground bounce from the higher clock frequency transitions may appear as spikes on the lower frequency clock signals (see Figure 9-1).

When advanced Schottky or advanced CMOS devices are used on boards of any significant size, wave shape must be controlled for clock lines. When actual routed line lengths are considered, clock-signal lines on most boards will have propagation delays that exceed one-half the rise time of the clock signal at its source. Thus, on most boards termination must be used to ensure wave-shape control. Board clock lines should be terminated so that they are slightly underdamped for optimum edge speed, but they must not be so underdamped that excessive ringing occurs. Underdamped signals have faster rise times and thus reach threshold levels faster than critical-damped or overdamped signals. A suggested range for termination impedance is 1.2 to 2 times the characteristic impedance Z_o of the line being terminated.

The clock line termination task on circuit boards differs from that of the other two categories. In contrast to unit-to-unit and board-to-board clock distribution, where clock interconnections can be arranged so that there is only one source and one load for each clock line, board clock signals must be routed to a large number of destinations. On a circuit board, it is impractical to limit clock lines to a single source and load; multiple loads must be driven by drivers in all but the most trivial designs. A typical circuit board is populated with numerous clocked devices, such as flip-flops, counters, registers, and shift-registers, all of which require clock signals. When synchronous design practices are followed, as they should be, all clocked devices on a board require a clock signal. Thus, in most cases, board-level clock distribution needs are such that a driver per load is not practical; thus the optimum configuration for terminating lines, one source and one load, is not possible at the board level.

In most applications it is impractical to load terminate each clock line on a board. Load termination dissipates a great deal of power, and split load termination increases system parts count and cannot be used with CMOS input threshold levels. There are schemes for ac load terminating lines that reduce power dissipation, but they require a larger number of discrete components.

Series or source termination using ferrite shield beads is the preferred method of board-level clock-signal line termination. Shield beads, which are made of a lossy ferrite material, present little impedance to clock signals at low frequencies, but at high frequencies they act as small lossy impedances that limit high-frequency energy. When a shield bead is used for source termination, the terminated line must pass through the bead and the bead must be located very near the driver (within 1 in). Beads with impedances in the range from a few ohms to near 50 Ω at 100 MHz are available. Clock driver edge rates and circuit board characteristic impedance should be considered when selecting a bead. When stitch-weld or wire-wrap boards, which tend to have a characteristic impedance on the order of 100 Ω, are used, beads with an impedance near 50 Ω at 100 MHz should be selected. For printed circuit boards with typical Z_o's near 50 Ω, a bead of 25 Ω at 100 MHz should be considered. The goal is to not overdampen clock signals, so that the edges are not slowed down too much. The idea is to remove enough energy to prevent excessive overshooting, which in turn will prevent undershooting of clock signals.

Small resistors can also be used for series termination of board clock lines. The disadvantage of series resistor termination is that the dc noise margin is reduced when driving TTL devices. However, the reduction is generally very minimal for advanced Schottky loads since most advanced Schottky devices have very low, *low*-level dc input currents (with the exception of certain devices such as F240, F241, and F244), but caution must be exercised. The noise margin of the clock lines can be compromised unless the situation is evaluated carefully. For example, the *low*-level input current specification for standard FAST devices is 0.6 mA so that the total dc load current for six loads is 3.6 mA. If a 51-Ω series resistor is used, the drop across the resistor, and the loss in *low*-level noise margin, is 183.6 mV (3.6 mA × 51 Ω). A loss of 183.6 mV is a significant portion of the 300-mV *low* noise margin; such a large reduction in noise margin cannot be tolerated in most applications. A 51-Ω resistor is unacceptable if the load consists of six older high-speed Schottky (S) devices or of FAST octal parts such as F240, F241, or F240. *Low* input current is 2 mA for S parts, 1 mA for F240, and 1.6 mA for F241 and F244. With six loads, the drop across a 51-Ω series-terminating resistor is 612 mV with 2 mA per input (2 mA × 6 × 51 Ω) and 306 mV with six 1-mA loads. The drop in either case exceeds *low*-level TTL noise margin, which is 300 mV. Thus, series resistors should not be used in such applications, or much lower value resistors must be used. Series termination is ideal for CMOS devices, since CMOS dc input current is very low—typically less than 1 μA.

Series termination, either with shield beads or resistors, has the advantage of requiring only one discrete component per line and does not dissipate any extra dc power. Series termination does present one problem; when long lines are driven, signal transitions near the source will step up (or down) as shown in Figure 9-10. Steps are inherent in signal (clock) transitions near sources when long lines are source-terminated (see Section 11-5-5); the result is that signals may dwell for some period of time in the threshold region of receiving devices. When a step dwells in the threshold region of a device, double clocking of the device may occur as a result of ringing or noise on the line operating in conjunction with the indeterminate logic level of the step. To avoid extraneous clocking, clock lines must be kept very short or loads must be grouped very close together at the end of lines (within 4 in is good).

Board clock circuitry must conform to the best possible electrical design standards. Each clock driver package should be decoupled with a high-frequency decoupling capacitor, such as a 0.1-μF ceramic capacitor, located as close to the package as possible and connected between

Figure 9-10   Inherent step near the source of a series-terminated line.

the power and ground pins of the package as directly as possible. The clock drivers themselves must be connected as directly as possible to the power and ground planes or grids of the host board. When stitch-weld or wire-wrap boards are used, clock driver package power and ground connections and decoupling capacitor connections must be made directly to the power and ground planes with solder clips or washers. Wiring should never be used for power or ground connections of clock circuitry; the inductance of wire connections aggravates ground-bounce problems.

When board-to-motherboard connectors do not contain an adequate number of ground pins, or when other noisy reference conditions exist, some form of balanced differential connection to the clock source must be used to ensure a clean, noise-free, board clock source. If standard TTL or CMOS interfaces are used when board-to-motherboard connectors do not contain an adequate number of ground pins, noise may be injected into the clock signals; noise cannot be tolerated on clock signals. Thus, if single-ended clock inputs are to be used, a low-impedance reference path must exist between board reference planes and the motherboard reference plane to prevent reference-level shifts in excess of single-ended TTL or CMOS noise margins when large load currents are switched.

Gating of board clocks should be avoided where possible because it places severe restrictions on the timing of the gating signals relative to the clock. It is more desirable to gate a clocked element's control signals instead of the clock itself. However, when gating is required, two-level clock trees, as shown in Figures 9-9 and 9-11, do provide a safe means of gating clocks. The intermediate level (CLK) can be safely gated with signals that originate from devices clocked with the final stage (CLK–) of the clock tree (see Figure 9-11). No spiking occurs on the gated clock line since the control signal safely brackets the intermediate clock signal (CLK in Figure 9-11). If boards are built with only one level of clock-signal buffers, there is no easy means of ensuring that gated signals have the proper time relationship to nongated clock signals (gated signals will lag) nor is there means of ensuring that gate control signals bracket clock signals. Thus, gated clocks tend to spike.

Summary of board-level clocking techniques

1. Use only a single clock frequency and phase.
2. Use a two-level clock tree.
3. Use the same type of devices (preferably devices in the same package) in each given level of a clock tree.
4. Balance clock-signal loads.
5. Centralize the distribution point for board clock lines.

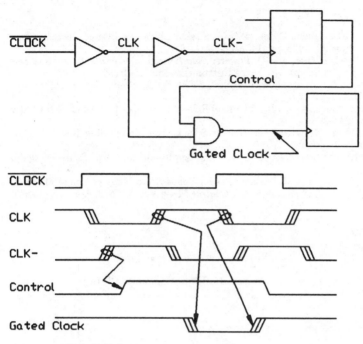

Figure 9-11   Gated clock.

6. Keep clock lines short or balance clock line lengths.

7. Limit the number of ac loads per clock driver package to keep ground bounce within limits.

8. Do not mix clock frequencies, phases, or other signals in clock driver packages.

9. Provide a decoupling capacitor for each clock driver package and locate it as near as possible to the package power pins.

10. Directly connect the power and ground pins of clock driver packages to board power and ground planes (no wiring).

11. Terminate all clock lines to ensure high-quality clock signals.

12. Closely group clock line loads near the end of clock lines to prevent false triggering due to edge steps near series-terminated board clock sources.

13. Isolate clock-signal lines to minimize cross coupling.

14. Use only those devices that clock on *low*-to-*high* transitions when using devices with TTL input levels.

15. Avoid gating of clocks where possible.

References

1. Royle, David: "Designer's Guide to Transmission Lines and Interconnections, Part Two," *EDN*, June 23, 1988, pp. 143–148.
2. Abramson, S., C. Hefner, and D. Powers: *Simultaneous Switching Evaluation and Testing Design Considerations*, Texas Instruments Inc., Dallas, Tex., 1987.
3. Cowdell, Robert B.: "Bypass and Feedthrough Filters," *Electronic Design*, No. 17, August 16, 1975, pp. 62–67.
4. Ormond, Tom: "Backplanes Play a Crucial Role in High-Speed Systems," *EDN*, July 10, 1986, pp. 222–228.
5. Visco, P. Anthony: "Coaxing Top Bipolar Speeds from Prototyping Boards," *Electronic Products*, September 1, 1987, pp. 55–58.
6. *FAST Applications Handbook 1987*, National Semiconductor Corp., South Portland, Me., 1988.
7. Parker, C., B. Tolen, and R. Parker: "Prayer Beads Solve Many of Your EMI Problems," *EMC Technology & Interference Control News*, Vol. 4, No. 2, April–July 1985.

Bibliography

Fletcher, William I.: *An Engineering Approach to Digital Design*, Prentice-Hall Inc., Englewood Cliffs, N. J., 1980.
Mano, Morris M.: *Digital Design*, Prentice-Hall Inc., Englewood Cliffs, N. J., 1984.

Signal Interconnections

Each distinct class of digital system or subsystem has specific interconnection requirements, some being much more critical than others. However, when advanced Schottky or advanced CMOS devices are used to implement a system, the interconnection system and the signal routing, at any of the system levels or boundaries, are always critical and cannot be left to chance or be based on purely mechanical concerns. Signal-path electrical requirements must be given high priority for system operating speed to approach that of the usable intrinsic logic device speed. Until recently, system operating speed was limited by the digital components, but that is no longer the case when high-speed advanced Schottky or advanced CMOS devices are used. The bandwidth of the interconnection system is often responsible for a major portion of the delays that limit system operating speed.[1]

10-1 Signal Interconnection Categories

Signal interconnections fall into three basic categories in most large digital systems:

1. Interconnections within a circuit card (component-to-component)

2. Interconnections between circuit cards that are mounted on a common motherboard (board-to-board)

3. Interconnections between separate units such as boxes, chassis, racks, cabinets, etc. (unit-to-unit)

Each category has specific physical and electrical design and implementation requirements. These special requirements must be understood so that the proper interface devices and physical interconnections can be selected. Fundamental issues that must be addressed at each interconnection level include signal dynamic response requirements, interconnection and device propagation delays, transmission-line effects, and noise.[2]

Noise is of particular concern for interconnections between physically separated units. The unit-to-unit interface circuitry must be able to operate properly in the presence of high levels of noise and dc reference offsets.

10-2   Physical Interconnections

Large high-speed digital systems typically use the following types of physical interconnections for the three interconnection categories[3–5]:

1. Multilayer printed circuit wiring boards for component-to-component connections

2. Multilayer printed circuit or wire-wrap backpanels for board-to-board connections

3. Cables composed of twisted-pair or shielded twisted-pair wire for unit-to-unit connections

In a low-speed system, most signals may not require much care in the design or selection of the interconnection media and routing, but most systems employing advanced Schottky or advanced CMOS devices are expected to operate near the limits of possible device or system performance. Hence, a great deal of design time is usually necessary to optimize the interconnect network for most signals, particularly clocks and critical data paths such as memory data or address lines.

The fast edge rates of advanced Schottky or advanced CMOS devices will cause severe cross coupling between signals, noise in the power and ground distribution systems, and degraded signals due to transmission-line effects (all of which will affect correct system operation) unless the physical interconnect system is designed to limit the side effects of fast edges. When devices have fast edges, the arrangement of the physical routing of all signals is critical.[6]

To minimize the undesirable side effects of fast edges, all lines must be kept as short as possible to minimize transmission-line effects. The length and spacing of parallel runs must be controlled to minimize cross coupling (see Chapter 11). Critical signals must be properly terminated to prevent transmission-line effects such as excessive overshoot or undershoot. To minimize reflections, critical lines should not branch, and the length of any stubs should not exceed one-fourth of the rise time of the driving signal.[7]

Multilayer pc boards are generally required for modern high-performance systems at the component-to-component level because of the large number of interconnections required by most designs. High signal density increases the probability of coupled noise. Thus, be sure to place components on circuit boards so as to minimize signal runs. Reducing signal line length reduces the capacitance load, minimizes the chance for crosstalk, and reduces transmission-line effects and signal interconnection propagation delays. The placement of components should never be left to the drafting or mechanical departments; the electrical designer must be responsible for component placement.

Advanced Schottky or advanced CMOS interconnection circuit boards must have continuous ground and voltage planes that are not broken or interrupted by device and connector pin fields (planes must be connected between the component pins). The ground plane must be arranged so that it is below (or above) all signal runs on the board, including signal traces in the connector pin fields. A reference plane placed near each signal trace provides a low-impedance return path for transient load currents and provides a shielding effect that reduces cross coupling between signals.

Board-to-board connections are typically implemented in one of the following three ways:

1. A multilayer pc motherboard with multiple signal layers, and ground and voltage planes

2. Wire-wrap signal connections above a multilayer backpanel with ground and voltage planes

3. Wire-wrap signal connections, bus bars for power connections, and a backpanel that serves as a ground plane

With attention to the basic concerns of a high-performance signal distribution system, any of the above structures can be used successfully for most advanced Schottky or advanced CMOS applications. Very high speed applications require multilayer mother-daughter pc boards with controlled-impedance signal layers (instead of standard printed wiring interconnects that do not have controlled impedances), so that the response of the interconnections can be accurately predicted and controlled. In all cases, daughterboard-to-motherboard connections must have a sufficient number of ground pins so that transient load currents do not cause significant shifts in the reference level of the daughterboards.[8,9] As a rule of thumb, daughterboard–to–motherboard connectors should have ground pins located at each 0.5- to 1-in increment of linear distance to ensure a low-impedance ac connection between daughterboards and motherboard reference planes.

At the board-to-board level, signal connections tend to be dominated by bus structures that have an increased probability of crosstalk because of the long runs across backpanels or motherboards. Make sure that daughterboard connector signal pins are assigned in such a way that the lengths of signal lines are minimized and that signals are not bundled together if they can be upset by cross coupling (techniques for controlling crosstalk are discussed in Chapter 11).

Breadboard interconnections. During the breadboard phase of most systems, welded-wire or wire-wrap interconnections may be used at both the component-to-component and the motherboard level to expedite the completion of the initial breadboard unit and to allow corrections or modifications to be easily incorporated.

When breadboarding with off-the-shelf welded-wire or wire-wrap boards, only those boards that are described as Schottky boards should be used when high-speed devices are being applied. Schottky boards have ground and voltage planes that are interconnected between all device and connector pins.

When advanced Schottky or advanced CMOS devices are interconnected with wired boards, the signal response characteristics will not be representative of the final pc board system. Thus, wired breadboards should only be used to establish the functionality of a logic design; they should never be used to establish the operating speed of the final pc board system. When the limits of the operating speed of an advanced Schottky system must be demonstrated, the breadboard system must have the same physical dimensions and electrical characteristics as the final product. The mechanical and electrical designs cannot be separated when advanced Schottky devices are used.

10-3   Electrical Interconnections

The characteristics needed for electrical interfaces are highly dependent on the stability of the ground reference system. If a low-impedance reference can be maintained, single-ended interconnections can be used. If a low-impedance reference cannot be maintained, differential communication is required. In the typical large system situation, it is possible to maintain a solid ground reference at the component-to-component and board-to-board level but not at the unit-to-unit level. The physical spacing between units in most large system situations makes it impractical to provide a sufficiently low-impedance reference path between units. Thus, the typical high-performance system built with advanced Schottky or advanced CMOS components uses the following electrical interfaces:

1. Single-ended TTL or CMOS levels for component-to-component and board-to-board communication

2. Balanced differential line drivers and receivers for unit-to-unit communication.

10-3-1   Component-to-component and board-to-board connections

Component-to-component connections on circuit boards and connections from board to board where there is a continuous common ground reference plane (motherboard with a ground plane) can be reliably transmitted and received using single-ended TTL or CMOS levels.

Most component-to-component and board-to-board connections are implemented with single-ended signal interconnections, which require the minimum possible number of interconnections (one). Single-ended driver-receivers are as simple as possible; thus single-ended communication minimizes complexity and cost. However, single-ended communication is only possible when the interconnected components are referenced to a common solid ground plane (see Figure 10-1). Thus, circuits and motherboards must have continuous ground planes with no cut-out areas, and circuit board–to–motherboard connectors must have a sufficient number of evenly distributed ground connections so that the circuit board ground plane appears as an extension of the motherboard ground plane.[8,9] Circuit board–to–motherboard ground connection inductance must be low enough so that transient return currents do not shift the reference level of circuit boards beyond single-ended threshold limits.

> Single-ended TTL or CMOS communication requires a solid common ground reference plane.

At the board-to-board level, signal connections tend to be dominated by bus structures that have large capacitance loads and an increased probability of crosstalk due to long runs across backpanels or motherboards. Thus, the effects of large loads and noise on signals must be taken into account in the design of interfaces that have signals that traverse motherboards.[10]

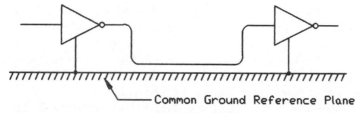

Common Ground Reference Plane

Figure 10-1   Single-ended signal interconnection.

10-3-2   Signal connections between remote units

At the unit-to-unit level, significant ac and dc differences in the reference (ground) potential of the various units will generally be present. Furthermore, signals that are routed between units are usually exposed to a more hostile noise environment than signals that remain internal to a given unit or subsystem. Thus, communication between separate units where the actual transmitting and receiving components are *not* directly referenced to a common reference plane, such as the ground plane of a board, motherboard, or backplane, requires a type of signal transmission that has a high level of noise tolerance. Standard single-ended TTL or CMOS devices do not provide enough noise immunity. Single-ended devices do not have enough noise margin to function reliably unless both the driving and the receiving devices are directly referenced to a continuous common ground plane, which is not possible for drivers and receivers that are in separate units. As shown in Chapter 5, the switching of one advanced Schottky TTL driver causes voltage transients across a 1-in length of wire that exceed TTL *low*-level noise margin (0.3 V). Thus, transient differences that exceed 0.3 V should be anticipated between units that require several feet of wire to interconnect.

Signals transmitted between remote devices (i.e., devices that are not tied directly to a continuous common ground plane) must be transmitted by some method that will function reliably in the presence of high levels of noise and offset voltages. Single-ended TTL devices are not adequate.

High-speed differential unit-to-unit signal transmission.   At present, most high-speed digital communication links between separate units (i.e., units not referenced to a common ground plane) use some form of balanced differential electrical interconnection. Most are configured to meet the requirements of the Electronic Industries Association (EIA) Standard RS-422.[11]

A balanced differential interface network consists of a differential line driver connected to a differential receiver by a twisted-pair, or a shielded twisted-pair, of wires as shown in Figure 10-2. Properly configured balanced two-wire differential communication networks are insensitive to noise or offset voltage common to both lines unless the voltage exceeds the operating range of the network.

Source or destination reference offsets, which often occur as the result of dc or ac current flow in the ground reference system, appear to

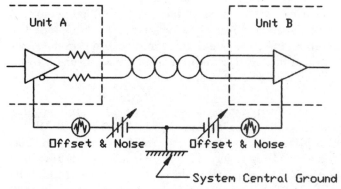

Figure 10-2   Balanced differential unit-to-unit interconnection.

be common to both lines of a differential interunit connection (see Figure 10-2). Also, noise coupled into twisted-pair lines generally appears equally on both lines of a differential connection.[12] Thus, interunit communication requires two-terminal line receivers that ignore common input levels. Line receivers with true differential inputs serve that purpose; true differential input stages are insensitive to noise and offset voltages common to both inputs (within their operating limits). Each data set is conveyed by the polarity of the difference in voltage between a line pair not by the absolute magnitude of the voltage. The ability to reject noise or offset voltages common to both inputs is referred to as common-mode noise rejection.[13] Differential receivers that have, for the application, a sufficient amount of common-mode noise and common-mode dc offset rejection capability are required for interunit communication.

To reject common-mode noise, the receiving circuit must have a differential (difference) input stage such as shown in Figure 10-3. A differential input amplifier stage is insensitive to the absolute input voltage. Each data set is conveyed by the polarity of the voltage difference between the two inputs not the absolute input voltage levels.[13]

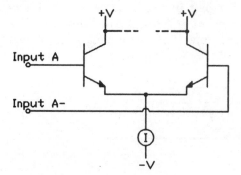

Figure 10-3 Differential input stage.

Balanced differential signal transmission has the added advantage of generating little noise in the ground system. The transient line voltages and currents tend to be equal and opposite and therefore cancel. Thus, balanced differential communication between units limits the noise pollution of the local environment.

Differential unit-to-unit drivers and receivers. For unit-to-unit applications where data transfer rates do not exceed 16 Mb/s (16 Mb/s is equivalent to an 8-MHz clock rate—see Figure 10-4) and the environmental conditions are not too severe, IC 26LS31 differential line drivers and 26LS32 receivers are possible device choices (see Figure 10-5).[14,15] However, 26LS31 drivers and 26LS32 receivers should not be used above 8 MHz; they may overheat and fail.

For applications with clock rates above 8 MHz, several new 422-compatible driver-receiver pairs are available, such as the 26C31 and 32, 26F31 and 32, and 75ALS192 and 193; 75ALS192 and 193 driver-receiver pairs will operate up to 20 MHz (clock rates), 26F31 and 32 up to 25 MHz, and 26C31 and 32 up to 40 MHz (the 26C31 and 32 are only available in the −40 to +85°C temperature range).

The various 422-compatible receivers have the ability to function correctly in the presence of ±7 V of common-mode noise or dc offset, which is adequate for most unit-to-unit applications. For very noisy applications, a modified version of the basic 26 XX 32, the 26 XX 33, with ±15 V of common-mode rejection is available. The 26 XX 33 has the same basic internal circuitry as the 26 XX 32, but the 26 XX 33 has a larger ratio input divider network which allows operation over a wider input-voltage range. The penalty is less sensitivity (i.e., the required minimum difference in the two input levels for a defined output), 0.5 versus 0.2 V for the 26 XX 32, but in many applications 0.5-V input sensitivity is more than adequate.

The various 26 XX 31 and 26 XX 32 drivers and receivers meet the requirements of the Electronic Industries Association (EIA) Standard RS-422 for data signal transmission between digital equipment.[11]

Twisted-pair or twisted-shielded–pair lines should be used to interconnect differential drivers and receivers. Twisted-pair lines minimize the effects of coupled noise, which tends to couple equally into each

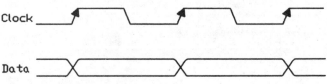

Figure 10-4   Clock and data waveforms.

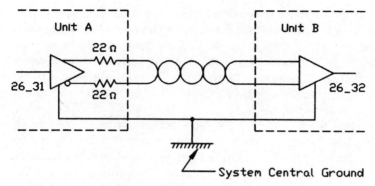

Figure 10-5   RS-422 differential line driver-receiver.

line of a line pair. Thus it appears as common-mode noise, which is rejected as long as it does not exceed the receiver common-mode operating limits.

Very high speed differential unit-to-unit drivers and receivers.   For communication at very high data rates between separate units (i.e., data rates above 80 MHz), there are no readily available, easily applied, off-the-shelf TTL- or CMOS-level drivers or receivers. High-speed analog comparators, such as the AM686 or the AD9686 with input divider networks to increase the common-mode range, are one solution when high-speed receivers are needed (see Figure 10-6).[16] Custom driver circuits, or TTL to ECL translators, such as the 10124/10524, are typically used for very high speed drivers.

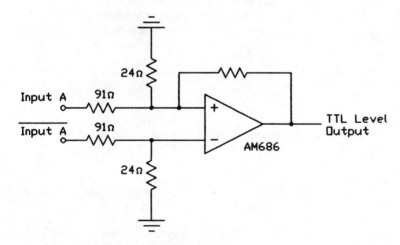

Figure 10-6   High-speed differential receiver.

Low-speed unit-to-unit signal transmission. For very low speed (less than 20 kb/s) data transmission between separate units, single-ended driver-receiver pairs (see Figure 10-7) that have signal levels that conform to the EIA Standard RS-232 are often used.[17] Single-ended RS-232 levels are considerably larger than TTL signal levels, and the greater the difference in digital signal levels, the greater the immunity to noise. Worst-case RS-232 *high* and *low* levels are +5 and −5 V. Typical levels are +9 and −9 V. A 10-V minimum logic level difference is sufficient to provide an acceptable amount of noise margin for most unit-to-unit applications. Large signal swings provide good noise immunity, but large signal swings do have disadvantages. It is difficult to switch quickly between levels; thus it is difficult to achieve high data rates. If high-speed switching is achieved, large, fast voltage transitions create a great deal of noise.

Each single-ended unit-to-unit signal should have an accompanying ground return line. Ground return lines are required to provide a direct low-impedance path for signal return currents. However, caution must be exercised when direct unit-to-unit ground connections are made.[18] Consideration must be given to the effects of interunit ground connections on overall system reference-level integrity. Large ground-loop currents may flow in the signal return connections, unless care has been taken in the overall system grounding to ensure that large offset voltages do not exist between units.[19]

Very low speed unit-to-unit signal transmission. Many digital systems must receive a large number of very low frequency (near dc) bilevel signals from mechanical switches, relays, solenoids, or other such external sources. Signals from such sources are generally very noisy. Yet, often little attention is given to the special interface needs of such signals. In many instances, receivers without adequate noise rejection

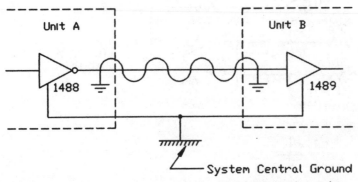

Figure 10-7   RS-232 low-frequency single-ended driver-receiver pair.

are used, with the apparent assumption being that noise is not a problem when signal frequency is low. External signals, even if their frequency is near dc, must be received in such a manner that unambiguous logic decisions can be made and that noise associated with the signals, or signal references, does not propagate into the receiving system. Cost considerations and the need for simplicity often dictate that such signals be received with single-ended receivers, which complicate the task, however. If single-ended communication is used for external interfaces, signal levels must be large enough so that it is easy to distinguish signals from noise, or signals must be heavily filtered to remove noise. Single-ended TTL levels are inadequate for noisy external sources not solidly referenced to the same ground as the receiver.

Very low frequency receivers. High input impedance CMOS logic buffers are ideal for many low-frequency receiver applications. The high intrinsic CMOS buffer input impedance provides isolation between source and load and allows the use of a number of filtering techniques that cannot be used with TTL buffers. Buffers with high input impedance allow the use of very high impedance input divider-filter networks (TTL buffer input current requirements severely restrict signal source impedance). For example, a CMOS buffer preceded by a divider-filter network, as shown in Figure 10-8, provides a simple, inexpensive means of receiving noisy, low-frequency high-level external signals and translating them to TTL or CMOS levels.

The input divider allows signals with level changes greater than the buffer threshold voltages to be reduced to levels compatible with the buffer input requirements, and if the divider ratio is adjusted, a large range of input levels can be received. For example, bilevel signals, such as those that might originate from 12- or 28-V relays or solenoids, can be translated to levels compatible with 5-V powered CMOS buffer input requirements. The maximum input level is only limited by practical considerations, such as limiting voltages in digital units to safe and reasonable levels. Minimum input signal levels must at

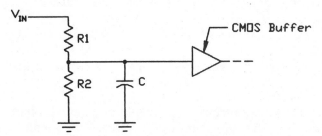

Figure 10-8   Very low speed high-level input buffer.

least exceed the thresholds limits of the receiving CMOS buffer. Both minimum and maximum *high* and *low* signal levels must be considered when selecting the divider network ratio to ensure that normal signal-level variations at the divider output are sufficient to ensure switching but do not significantly exceed V_{cc} or ground. However, if the divider input resistor $R1$ is large enough, V_{cc} or ground can be exceeded without harm to the buffer. The resistance of $R1$ must be such that under all overvoltage or transient conditions the current injected into the CMOS buffer input protection circuit is less than the specified limits for the device. High-impedance input divider-filter networks also serve a number of other purposes, including

1. Limiting the loading of the source

2. Limiting circulating ground currents

3. Limiting fault or transient input currents injected into the accompanying CMOS buffers to safe levels[20]

4. Allowing heavy filtering with small-value capacitors, and thus physically small capacitors

Since filtering will slow the rise time of signals at the buffer input, buffers with minimum rise-time requirements cannot be used in such applications. Devices that can tolerate slow rise-time inputs, such as 4009 or 4010, or 4049 or 4050 level translator buffers, or 54HC14 or 54AC14 Schmitt trigger buffers, are suitable for such applications. Most high-speed (HC) or advanced CMOS (AC) devices cannot be used in such applications since most CMOS devices have minimum rise-time requirements.

Unit-to-unit line terminations.   Series (source) termination should be used for most unit-to-unit interconnections.[21] Series termination provides short-circuit protection and does not increase static power dissipation as load termination does. For optimum response, the value of the terminating resistors (plus the output impedance of the driver) should be chosen so that the line is slightly underdamped, but not too much. A line that is slightly underdamped will provide a signal at the load that reaches the threshold region of the receiver more quickly than a line that is exactly matched or overdamped. The key to successful operation and optimum performance is to not underdampen the line so much that it rings back into the threshold region of the receiving device.

Miscellaneous unit-to-unit considerations.   All unit-to-unit signal cables must include ground lines for signal return currents. When single-ended high-level interunit communication is used (low-level single-ended signals should never be used), there should be one ground line per signal line. When balanced differential interunit communication is used, one ground line per four to eight signal pairs is a good rule of thumb. If differential signals are perfectly balanced, no return ground lines are needed, but differential signals are never perfectly balanced. Thus, some grounds are needed to provide a direct return path for the unbalanced portion of the signal currents. However, direct unit-to-unit ground connections offer the potential for large ground-loop currents, as shown in Figure 10-9. Thus, when unit-to-unit signal grounds are required to ensure a low-impedance path for signal return currents, make sure that all units have low-impedance ground connections to prevent large interunit reference voltage offsets and large ground-loop currents. If large offset voltages exist between units, ground current in the signal return lines can be of such a magnitude as to disrupt system operation by introducing noise and in severe cases overheat and burn out the signal return lines.

Optoisolators and fiber-optic links offer the promise of a convenient and effective solution for interunit communication. Optoisolators eliminate unit-to-unit ground-loop currents and eliminate reference offset concerns. However, they have not found widespread use at this time. Economic considerations and lack of compatibility with familiar electrical wiring techniques have delayed their acceptance.

Communication between remote units generally must incorporate some form of asynchronous data-transfer technique since clock alignment between remote units is usually difficult to maintain.

A note of caution. Sending TTL-level signals true-comp (comp = complementary) and receiving them with cross-coupled standard

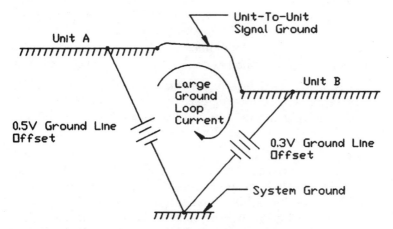

Figure 10-9 Ground-loop current flow in unit-to-unit ground lines.

single-ended TTL devices serves no useful purpose. Differential signal line receivers must have "true" differential inputs to reject common-mode noise and offsets.

10-4   Miscellaneous Signal Interconnection Considerations

The following interconnection design practices will help in the quest for achieving a solid reliable high-performance design (also see References 22 and 23):

1. Input signals to boards, racks, or systems should be buffered at one place so as to present only one load to the external world to minimize ac loading (minimizing ac loading is essential in high-speed systems).

2. Single-ended signals that traverse noisy areas, such as motherboards, should be received by devices with enhanced noise margin such as 54 XX 240, 54 XX 244, which have hysteresis and more noise margin than standard gates, or by Schimdt trigger buffers, such as 54 XX 14 or 54 XX 132, which have a high level of noise immunity.

3. Asynchronous input signals should never be run to multiple clocked devices within a functional unit; some of the devices within the unit may sense (sample) the signal at a slightly different time because of different line lengths or clock phasing at the multiple receiving points, and the unit may react incorrectly.

4. When asynchronous parallel buses must be received and synchronized in a unit, it is best to capture and synchronize the bus control or strobe signals rather than try to "broadside" synchronize the entire bus. There is no assurance that all of the signals on a wide bus will be properly captured when the data happens to be changing at the same time that the active clock edge occurs on the receiving storage elements.

5. Signals that are used internally and externally to a board or unit should be buffered before being used externally to isolate the internal signals from transmission line reflections or other external disturbances, such as shorts, that might occur external to the unit.

6. When devices with TTL levels are used, control signals that traverse noisy areas, such as motherboards, should have their logic polarity configured so that the predominant state (inactive state) is *high*. A *high* TTL level has a higher noise margin than a *low* level. If a signal has a state that is tolerant of noise or glitches, the signal should be logically arranged to be *low* for "the noise tolerant

state" since the *low* state has less noise margin (an asynchronous system reset is a possible example of such a signal).

7. The output of unbuffered clocked devices (registers, counters, flip-flops, etc.) should be buffered before being sent out of (or off) a unit (boards, boxes, etc.). Line reflections have been known to upset and change the output state of unbuffered clocked elements. Most advanced Schottky and advanced CMOS clocked elements have built-in output buffers, but many of the older logic families do not. Devices with three-state outputs have built-in buffers due to normal three-state buffer implementation techniques.

8. Signals routed to test points should be buffered to prevent test equipment from interfering with normal system operation. In particular, a buffered clock source should be provided for logic analyzers or other test devices so that the test clock can be appropriately terminated and so that long test lines do not corrupt internal clock signals.

9. Three-state signals (buses) should not be allowed to float when unused. Floating lines tend to drift into the threshold region of the receiving circuits, and noise or other disturbances that are coupled into the line tend to cause the input structures of the receiving devices to oscillate. The oscillation tends to increase the overall system noise, particularly on the power and ground distribution system; often the noise will reach unacceptable levels. Acceptable solutions are (1) to always drive three-state lines or (2) to provide pull-up resistors to ensure that lines do not float near the threshold level of the receiving devices when they are not being driven. The disadvantages of pull-up resistors include (1) they require added parts and (2) they cause increased power dissipation. Selecting a pull-up resistor value is often difficult; a trade-off between power dissipation and rise time of the line when not actively driven (i.e., when it is being pulled up) must be made. The use of pull-ups that are too large in order to reduce power dissipation may do more harm than good. Large-value pull-up resistors on lines with large capacitance loads may allow lines to remain in the critical noise generating area near the threshold region of the receiving devices for long periods. Furthermore, large-value pull-ups may pull signals into the threshold region that might have remained near their last active level if not influenced by a pull-up. When power dissipation is of concern and low-value pull-ups cannot be used, it may be best to always drive three-state lines *high* before releasing the lines (letting the drivers go into a high-impedance state). If three-state lines are driven *high*, high-value pull-up resistors can be used to hold three-state lines *high*.

10. Three-state drivers on a common line should be controlled in such a manner that they can never all (or more than one) be turned

ON at the same time, even for short periods during power turn ON or OFF. The safest method of achieving this is to use a single hardware decoder as the source of the enables for all drivers connected to a common line; then only one driver can be ON at one time (unless there is some fault). When it is not practical to provide a single decoder for the source of all line driver enables (for a common line or bus), which is often the case in large systems with bus interconnects that interface to numerous units, the system dc reset signal should be used to ensure that the controls to three-state drivers are initialized to the OFF state during power-supply transients and at turn ON. A dc reset should always be used for initialization. Requiring the presence of the system clock, which may not always be present because of faults or other reasons, for correct initialization of three-state buffers is an invitation for disaster. Most of the advanced Schottky or advanced CMOS three-state buffers are not rated for continuous short-circuit conditions.

11. Three-state bus systems should be designed so that the removal of a board or unit will not result in more than one set of the remaining drivers being turned ON.

References

1. Balph, Thomas: "Implementing High Speed Logic on Printed Circuit Boards," in 81 WESCON Session 18 Rec., Paper No. 1, September 1981.
2. Cutler, Robert: "Your Logic Simulation Is Only as Good as Your Board Layout," *VLSI System Design,* July 1987, pp. 40–42.
3. Crouch, Ronald A.: "Choose Cable with Care to Optimize System Design," *EDN,* November 5, 1978, pp. 113–116.
4. Timmons, Frank: "Wire or Cable Has Many Faces, Know Them All before Choosing, Part II," *EDN,* March 1, 1970, pp. 49–55.
5. Harper, Charles A.: *Handbook of Wiring, Cabling and Interconnections for Electronics,* McGraw-Hill, New York, 1972.
6. Royle, David: "Designer's Guide to Transmission Lines and Interconnections, Part One," *EDN,* June 23, 1988, pp. 131–136.
7. Royle, David: "Designer's Guide to Transmission Lines and Interconnections, Part Two," *EDN,* June 23, 1988, pp. 143–148.
8. DiCerto, Joseph: "Poor Packaging Produces Problems," *The Electronic Engineer,* September 1970, pp. 91–93.
9. Southard, Robert K.: "High-Speed Signal Pathways from Board to Board," in 81 WESCON Session 18 Rec., Paper No. 2, September 1981.
10. Royle, David: "Designer's Guide to Transmission Lines and Interconnections, Part Three," *EDN,* June 23, 1988, pp. 155–159.
11. EIA Standard RS-422-A, *Electrical Characteristics of Balanced Voltage Digital Interface Circuits,* Electronic Industries Association, Washington, D.C., August 1978.
12. Harper, Charles A.: *Handbook of Electronic Packaging,* McGraw-Hill, New York, 1969.
13. Millman, Jacob: *Microelectronics, Digital and Analog Circuits and Systems,* McGraw-Hill, 1979.
14. Cormier, Denny: "Serial Datacomm Driver/Receivers ICs Furnish Higher Data Rates, Lower Power Consumption," *EDN,* January 23, 1986, pp. 93–97.
15. Laws, David A., and Roy J. Levy: "Use of the Am26LS29, 30, 31 and 32 Quad Driver/Receiver Family in EIA RS-422 and 423 Applications," *Bipolar Microproces-*

sor Logic Interface Data Book, Advanced Micro Devices Inc., Sunnyvale, Calif., 1981.

16. "Technology Update, Representative Voltage Comparators," *EDN,* February 6, 1986, pp. 47–50.

17. EIA Standard RS-232-C, *Interface between Data Terminal Equipment and Data Communication Equipment Employing Serial Binary Data Interchange,* Electronic Industries Association, Washington, D.C., August 1969.

18. Brown, H. C.: "Get Rid of Ground-Loop Noise," *Electronic Design,* No. 15, July 19, 1969, pp. 84–87.

19. Oates, Edward R.: "Good Grounding and Shielding Practices," *Electronic Design,* No. 1, January 4, 1977, pp. 110–112.

20. Feulner, R. J.: "Solving Noise Problems in Digital Systems," *EEE, The Magazine of Circuit Design Engineering,* September 1967, pp. 79–83.

21. *Linear and Interface Data Book,* Advanced Micro Devices Inc., Sunnyvale, Calif., 1977, p. 4-21.

22. Sokal, Nathan O.: "Designer's Guide to PC-Board Logic Design—Part 1," *EDN,* November 13, 1986, pp. 253–262.

23. Sokal, Nathan O.: "Designer's Guide to PC-Board Logic Design—Part 2," *EDN,* November 27, 1986, pp. 229–235.

Signal Quality

High-performance advanced Schottky and advanced CMOS systems require high-quality noise-free signals that quickly stabilize. High-quality signals are only possible when a great deal of attention is given to the selection and configuration of the interface circuitry and the design of the interconnection system. Signal quality is a function of

1. *The ac loading:* Excessive ac loading slows data transfers.
2. *The dc loading:* Excessive dc loading may degrade signal levels and cause logic errors.
3. *Crosstalk:* Crosstalk may cause logic errors or require that additional time be allocated for crosstalk to subside.
4. *Transmission-line effects:* Transmission-line effects may cause logic errors and may require that additional time be allotted to allow signals to stabilize.

All of these factors must be considered over all of the worst-case conditions that a system must operate under.

11-1 Signal Response

As system operating speed increases, the delays in the interconnection system become a significant portion of the overall signal delays. Thus, for the maximum usable advanced Schottky or advanced CMOS device speed to be approached at the system level, the response of the interconnection system must be optimized. All interconnection delays must be minimized, signal lines must be kept as short as possible, and steps must be taken to ensure that ringing and other disturbances quickly subside.

There are several device parameters that are significant with respect to dynamic signal and system performance. They include the following:

1. Output rise and fall times (minimum and maximum)

2. Output current-voltage drive capability (both ac and dc)

3. Switching transient currents (internal and load related)

4. Ground upset voltage tolerance

5. Input threshold and noise tolerance

6. Input wave-shape requirements

Output rise or fall time is a critical parameter; faster edge rates increase crosstalk, transmission-line ringing, power-supply transients, and ground upset. Yet, minimum rise or fall times, which have the potential for causing the most trouble, are generally not specified for TTL or CMOS devices, particularly for lightly loaded situations. Lightly loaded advanced Schottky and advanced CMOS devices may exhibit rise times as short as 1 ns.

A knowledge of the dynamic response, which is a function of the current-voltage drive capability of the devices being applied, is essential if a high-performance system is to be realized. Yet, the dynamic drive capability of TTL or CMOS devices is not well specified. Worst-case dynamic limits are seldom specified, and typical dynamic (ac) drive values are often only shown on a generic basis in the logic family introduction that sometimes can be found in the manufacturers' data books. Hence, in most cases, worst-case dynamic drive characteristics must be estimated. The derating factor used for estimating the worst-case device parameter variations must account for the influence of process, temperature, and supply-voltage variations on device characteristics. A good rule of thumb is to add an additional ±50 percent to typical +25°C dynamic parameter values for most 0 to +70°C applications and ±100 percent for −55 to +125°C applications.

Switching transient currents, ground upset tolerance, and input threshold and noise tolerance are all interrelated. Internal and load-related transient current considerations are discussed in Chapters 5 and 6. Noise margin (noise tolerance) is discussed in Chapter 4. Advanced CMOS devices have higher noise margins than TTL devices and thus are less likely to be upset by a given level of transient switching current. However, advanced CMOS devices have larger output voltage swings and tend to be faster than TTL devices. Thus, advanced CMOS devices generate larger transient currents which in turn cause larger voltage disturbances than equivalent TTL devices. System designers, using standard components in standard packages, have no freedom to change inherent device specifications. However, system designers can, and must, ensure that system signal, power, and ground interconnections are designed to accommodate the devices being applied.

Advanced Schottky devices do not have input wave-shape requirements per se; there are no minimum or maximum input transition time limits (as there are with some CMOS devices). Clocks, or other edge-sensitive inputs, require clean, fast transitions, but TTL parts will not be damaged if the input signals are slow in transitioning. Most CMOS parts have maximum input transition time requirements. Slow input transitions allow both pull-up and pull-down complementary input FETs to be ON for some time, which can result in prolonged excessive V_{cc} to ground current flow and overheating.

The edge speeds of advanced Schottky or advanced CMOS devices require that clock signal or other edge-sensitive signals be terminated to achieve the needed wave-shape control. Unterminated advanced Schottky or advanced CMOS signals may ring, depending upon line length and other physical characteristics of the interconnection system and thus impact the system operating speed by requiring additional time for unterminated signals to settle (see Section 11-5).

In addition to the above-listed device dynamic considerations, the local power-supply decoupling required to support the maximum switching rates of the device into the worst-case transmission-line loading must be understood and provided (see Chapter 7).

11-2 Dynamic Loads

Dynamic loads fall into two basic categories; lumped capacitance loads and transmission-line loads, which are in effect distributed capacitance loads.

11-2-1 Capacitance loading

Excessive capacitive loading on the output of advanced Schottky or advanced CMOS devices can severely degrade the expected performance. The dynamic response of most advanced Schottky and advanced CMOS devices is specified with 50-pF loads. In contrast, most of the older logic families are specified with 15-pF loads; 50 pF is a much more realistic load. However, in many situations, actual loads will exceed 50 pF, particularly in many bus driving applications. In those cases where loads exceed 50 pF, some adjustment must be made to the specified output response time of the driving device. In some cases, device manufacturers provide data showing output response versus load capacitance. Figure 11-1 illustrates some typical data.

Figure 11-1 shows *typical* increases in output propagation delay for standard devices (not drivers) for the LS, S, F, and AC logic families versus load capacitance. Note that the data are typical data for nominal room temperature and supply voltage conditions. Worst-case data for response versus load capacitance are seldom provided by device manufacturers. Multiplying typical values by a factor of 2 provides a reasonable estimate of worst-case values. The factor of 2 is based on possible device characteristic changes due to process variations and operation at the limits of the temperature and supply-voltage range.

When delay versus load capacitance information is not provided, an estimate of the increase delay due to larger-than-specified load capacitance can be determined by

$$\frac{dv}{dt} = \frac{I_{\text{SC}}}{C} \qquad (11\text{-}1)$$

where I_{SC} is the output short-circuit current of the output driving device and C is the increase in load capacitance beyond the specified load value. For TTL devices, or CMOS devices with TTL output levels, the value of I_{SC} is different for *high* and *low* outputs. In most cases, *low*-to-*high* transition times are longer than those of *high*-to-*low* transitions. Most TTL devices have lower *high*-level I_{SC} than *low*-level I_{SC} because of the current-limiting resistors used in TTL output pull-up circuits. Most CMOS devices that have TTL output levels use *n*-channel source followers for output pull-ups; such circuits tend to have less drive than conventional switched CMOS outputs.

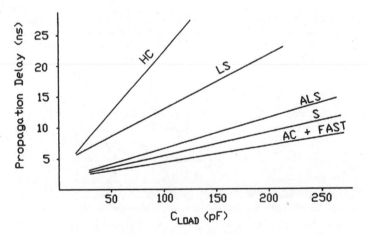

Figure 11-1   Output response versus load capacitance.

11-2-2   Transmission-line loads

In most applications where advanced Schottky or advanced CMOS devices are driving *long* lines (see Section 11-5 for the definition of a *long* line), the dynamic load should be viewed as a transmission line.[1,2] When the line is long, it is incorrect to treat the total line capacitance as a lumped load.[3] The line impedance isolates the driver from the more remote capacitance of the line and loads. Thus, in long-line cases the dynamic load is the characteristic impedance Z_o of the line, not the total capacitance of the line. In *long*-line cases, the Z_o of the line should be used to calculate the transient driver and load currents and transmission-line analysis must be used to determine the transient signal response (see Section 11-5).

It is desirable to have interconnection impedances as high as possible to reduce transient switching currents, but high line impedance increases the chance for crosstalk. Thus, no ideal circuit board or motherboard interconnection impedance exists. Attempts to resolve the conflicting requirements indicates that line impedance near 60 Ω provides the optimal balance between switching current levels and crosstalk. However, most manufacturers of high-density multilayer pc boards have difficulty achieving line impedances of 60 Ω or greater. Thus, when pc boards are used, the main concern is one of ensuring that the line impedance is high enough so that signal line impedance does not cause excessive dynamic loading. To limit TTL and CMOS signal degradation, unloaded signal line impedance should be greater than 40 Ω (see Section 11-5).

11-3   Static Loading Limits

For a reliable design, never exceed the manufacturer's specified dc load limits, and for critical military and space applications, the manufacturer's ratings should derated somewhat. The manufacturer's dc load ratings for the logic families of most interest are listed in Chapter 2. Static (dc) loading calculations are straightforward, as long as only logic devices of the same family are being driven, and the limits are usually only exceeded as a result of carelessness. When logic families are mixed, when discrete networks are driven, or when line terminations are driven, make sure that dc ratings are not exceeded. Take special care when LS and HC or HCT parts are used in mixed logic family designs since LS or HC parts have very little dc drive (only 4 mA for most standard parts). For the advanced Schottky or the advanced CMOS logic families, dc drive is seldom a limiting factor; ac (dynamic) response is usually responsible for limiting system operation. When a system built with advanced Schottky or advanced CMOS devices is found to be dc drive limited, the design should be carefully reevaluated. The capacitance of a signal line that runs to enough devices to fully dc load an advanced TTL or CMOS device will severely restrict that device's dynamic performance.

Applications that need high reliability usually require that some derating factor be applied to manufacturers' dc drive ratings. For military systems, dc drive is usually limited to 70 or 80 percent of the manufacturer's ratings. For space applications, 50 percent derating is often used. Most government contracts specify a required dc derating. Derating of dc drive provides some safety factor to compensate for weak parts that escape the screening process, for degradation with age, and for the fact that the actual circuit in which a part is used is seldom identical to the circuit and load used to test the device. Since there is no assurance that a part will perform exactly as indicated in the specification when it is used in a different (from the test) circuit configuration, dc drive derating is needed for high-reliability applications.

11-4   Crosstalk

Crosstalk between signals occurs as a result of capacitive and inductive coupling between adjacent or nearby lines. The faster (relative to older logic families) edge rates of advanced Schottky and advanced CMOS devices greatly increase the possibility of crosstalk between signs.[4] Increased coupling plus faster device response greatly increase the possibility that system operation will be degraded by crosstalk. Synchronous design practices reduce the possibility of crosstalk disrupting system operation (see Chapter 8), but when advanced Schottky or advanced CMOS devices are applied, the goal is usually to achieve high operating speed. Thus, there may be little time for the crosstalk that may occur to subside.

Crosstalk is a function of the separation between signal lines, the linear distance that signals lines run parallel with each other, and the height above a ground or other reference plane. Thus, where coupling between signals is of great concern, signals should be run at right angles to each other and kept as close as possible to a ground plane or isolated from one another by ground traces and reference planes. For example, signals such as memory (RAMs, ROMs), address, and data lines, should be run at right angles to each other to minimize cross coupling. Excessive cross coupling between memory device data and address lines can cause feedback that can affect the response time of the memory device and in extreme cases result in unstable oscillatory outputs.

Multilayer pc boards simplify the task of providing isolation; critical signals can be routed at right angles on different layers, and voltage and ground planes can serve as isolation layers. Where proper sys-

Figure 11-2 Twist signal with ground for critical single-ended signals.

tem operation may be highly susceptible to crosstalk, the signals of concern should be isolated between reference planes on multilayer pc boards. All clock signals should be isolated from other signals by reference planes. Critical signals can be further protected by running a ground trace on each side of the critical signal trace. A noisy signal should not be placed directly above or below a critical signal that is not isolated by a reference plane.

On welded-wire or wire-wrap boards or backpanels, the wiring should be as direct as possible between points so as to randomize the routing and be kept as close as possible to the ground (or voltage) plane. Make sure that the wiring is not channelized; this is especially important for critical signals such as clocks, so that they do not get channelized with noisy signals.

On welded-wire or wire-wrap boards or backpanels, critical single-ended signals should be run twisted with a ground wire connected to ground at the source and at the load as shown in Figure 11-2.

Twisting signal lines with ground lines provides the most direct path possible for return currents and provides some shielding effect. Return currents tend to flow in the nearby twisted ground line and not in the power or ground planes (see Figure 11-3). Thus, transient current flow and noise are reduced in the power and ground system. In addition, twisted-pair signal lines provide better control of line impedance, which allows more accurate terminations which in turn helps control ringing and line reflections. Twisted-pair lines also tend to confine the magnetic fields of the two twisted conductors, which minimizes the chance for coupling into adjacent wiring.

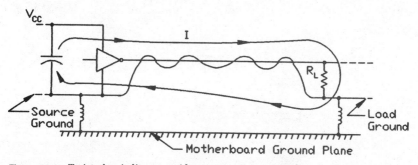

Figure 11-3 Twisted-pair lines provide a return current path.

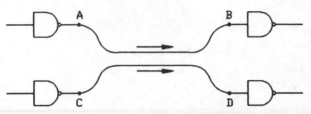

Figure 11-4   Coupled lines with the same signal flow.

Most high-performance advanced Schottky and advanced CMOS systems require a large number of circuit board and motherboard interconnections. Where signal interconnection needs force high signal density, some signal line must inevitability run close to and parallel with other signals. In such cases, two forms of crosstalk exist—forward and backward. Forward crosstalk is present on coupled lines in time coincidence with an active wave front on a nearby driven line and exists for the duration of the edge transition of the driven line. Backward crosstalk on coupled lines flows away from the wave front on the nearby active line and exists for twice the propagation delay of the coupled line length.[5] Both forms of crosstalk can cause circuit malfunction. Backward crosstalk tends to be more detrimental, since it tends to be of a higher amplitude and to last for a longer time, but either form can cause logic errors. The extent of the possible upset of a coupled-to line is dependent upon the polarity and amplitude of the coupled noise relative to the logic level of the signal that is disturbed and the physical topology of the lines. The signal flow may be such that the coupling is of little concern. For example, backward crosstalk is of no concern at output node C in Figure 11-4. However, backward crosstalk could be a major problem for a circuit with the topology shown in Figure 11-5.

In the case shown in Figure 11-5, backward crosstalk may or may not cause a problem depending upon the polarity of the coupled signal. If the coupling adds to or subtracts from the existing level so that the resulting composite signal has more margin, then the coupling is of

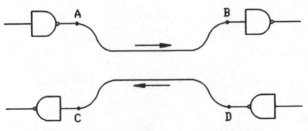

Figure 11-5   Coupled lines with reverse signal flow.

little concern (assuming that the coupled signal does not cause excessive ringing). Since it may be impractical to analyze all possible combinations, the safest approach is to assume that there will be some combination of active and inactive signals, such that some inactive signals will be degraded by crosstalk.

In those cases where the interconnect topology and the polarity of the coupling is in the harmful direction, the crosstalk may or may not be harmful depending on the amplitude and the duration of the coupled voltage. If the amplitude of the coupled voltage is less than the noise margin of the logic components used, then the coupling may not be detrimental. Likewise, if the duration of the coupled pulse is short enough, the receiving devices may not react to a narrow pulse even though the pulse exceeds the noise margin of the receiving devices. Some device manufacturers are specifying the pulse immunity of their devices to help designers evaluate the possible detrimental effects of ground-bounce transients. Such data are also useful for determining the sensitivity of devices to coupled pulses.

The duration of backward crosstalk is equal to the two-way delay of the coupled length t_l. Backward crosstalk reaches a maximum level when the propagation delay of the coupled length t_p is equal to one-half the rise time t_r measured from 10 to 90 percent of the active (driving) signal. Additional coupled length increases the duration of the coupled voltage but not the amplitude.

Forward crosstalk consists of a pulse with a width equal to the rise time t_r of the driving source. The amplitude is proportional to the coupled length.

The general expressions for crosstalk voltage amplitude between two lines (see Figure 11-6) for the two types of crosstalk are:[6,7]

Backward Crosstalk V_B

$$V_B = \left(\frac{K_C + K_L}{4}\right)\left(\frac{2t_p}{t_r}\right)(\Delta V_S) \qquad (11\text{-}2)$$

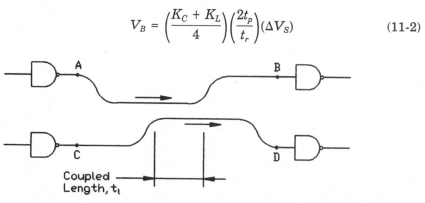

Figure 11-6 Coupled lines.

for coupled line lengths from $l = 0$ to $l = t_r/2t_{pd}$ and

$$V_B = \left(\frac{K_C + K_L}{4}\right)(\Delta V_S) \qquad (11\text{-}3)$$

for coupled line lengths of $l = t_r/2t_{pd}$ or greater.

Forward Crosstalk V_F

$$V_F = \left(\frac{K_C - K_L}{2}\right)\left(\frac{t_p}{t_r}\right)(\Delta V_S) \qquad (11\text{-}4)$$

For Equations (11-2), (11-3), and (11-4)

ΔV_S	driving signal transition amplitude
K_C	capacitive coupling coefficient
K_L	inductive coupling coefficient
t_{pd}	intrinsic propagation delay of the media
t_p	propagation delay of coupled length ($l \times t_{pd}$)
t_r	rise time of the driving signal
l	coupled length

and

$$K_C = \frac{C_m}{C}$$

$$K_L = \frac{L_m}{L}$$

where C_m = mutual capacitance between lines
$\quad\quad C$ = capacitance between lines and ground
$\quad\quad L_m$ = mutual inductance between lines
$\quad\quad L$ = inductance of each line

In a homogeneous material K_C and K_L are equal and no forward crosstalk exists.[7] However, conductors in typical digital applications are not surrounded by a pure homogeneous material. Welded-wire or wire-wrap interconnections, as well as pc board interconnections, are surrounded by a conglomerate of materials, other conductors, insulation, etc. Yet, for practical engineering purposes forward crosstalk is of little concern in welded-wire or wire-wrap circuit board or motherboard interconnections or in embedded conductors in multilayer pc boards. The exception occurs when two lines have a long coupled length (relative to the active signal rise time); in those cases the t_p/t_r term in the forward crosstalk equation becomes significant.

Backward crosstalk reaches a maximum value at a certain line length and does not increase in amplitude for longer coupled lengths. The line length where maximum amplitude is reached is related to rise or fall time of the driving signal. The limiting value is reached when the propagation delay t_p of the line is equal to one-half the rise time t_r of the active signal, i.e., at *the critical signal length*.[4] Thus, the coupled line length needed for maximum amplitude coupling is reduced as signal rise time decreases. A signal trace adjacent to a trace driven with a signal by a 2-ns edge transition requires one-half the coupled length to achieve the limiting value of crosstalk as is required for a signal with a 4-ns edge transition. Therefore, systems with fast rise-time signals will have significantly more crosstalk than systems with slow rise-time signals.

To calculate the magnitude of crosstalk, you must determine the line capacitance, inductance, and propagation delay, as well as the mutual capacitance and inductance between lines. All are difficult to determine. Actual measurements are the most reliable source for determining crosstalk, but the practicing engineer seldom has the time or resources to make controlled measurements on final production configuration boards. Yet, control of crosstalk, which requires a knowledge of its possible magnitude, is essential in high-speed systems. Otherwise, crosstalk will disrupt system operation. References 8 to 12 are recommended for assistance in predicting crosstalk. The DeFalco papers (References 9 and 10) are recommended as a starting point.

11-5 Transmission-Line Effects

Transmission-line effects at the pc board or motherboard level were of little concern with the older, slower logic families. The slow edge rates of early TTL or 4000 series CMOS devices (5 ns or greater) did not induce transmission-line effects, except on very long lines such as might exist between remote units. However, when advanced Schottky or advanced CMOS devices with edge rates of 2 ns or more are used, transmission-line effects are of concern at the pc board level. For all practical purposes, all signal lines are transmission lines when advanced Schottky or advanced CMOS logic components are used to implement digital designs.

Transmission-line effects (line response characteristics) are of concern, when signal runs (the length from the driving point to the most remote load) are equal to or exceed one-half the signal rise time t_r divided by the loaded (actual) propagation time t_{pd}' of the signal through the conducting media.[4,13,14] That is,

$$\text{Critical line length} = \frac{1}{2}\left(\frac{t_r}{t_{pd}'}\right) \tag{11-5}$$

The propagation delay for a typical pc board is 2 ns/ft. For a signal with a rise time of 2 ns, the *critical line length* is

$$\text{Critical line length} = \frac{1}{2}\left(\frac{2\text{ ns}}{2\text{ ns/ft}}\right) \tag{11-6}$$

$$\text{Critical line length} = 0.5\text{ ft} \tag{11-7}$$

Thus, in most applications where advanced Schottky or advanced CMOS devices are used, transmission-line effects must be considered since most signal runs, even on small pc boards, exceed 0.5 ft.

11-5-1 Basic transmission-line considerations

The following discussion is meant to convey a basic introduction to transmission-line theory and its application as they relate to high-speed TTL and CMOS devices and systems. A thorough discussion of transmission lines is beyond the scope of this book. A number of good references on the subject of transmission lines exist (see reference list at the end of the chapter), and all designers involved in the application of advanced Schottky or advanced CMOS logic devices should quickly procure (if they do not already have) some of the listed reference materials and undertake a thorough study of transmission lines if they are not currently conversant on the subject. A basic understanding of transmission lines and transmission-line effects is essential to the successful application of high-speed logic devices.

Signals transmitted on a line are well behaved when the signal source impedance, signal line impedance, and load impedance are very closely matched (i.e., almost equal) and when the line has only one source and one load with no discontinuities, branches, or stubs. In such cases, one line propagation delay (approximately 2 ns/ft if the line is in a pc board) after a signal is launched into a line at the source, a signal very much like the signal launched at the source can be observed at the load. If the signal source, line, and load are not matched, or the circuit topology does not consist of the ideal case of one source and one load, the signal quality will degrade as a result of transmission-line ringing and other effects. Signal sources, loads, and lines are not matched on a typical pc board, or on a welded-wire or wire-wrap logic device interconnection circuit board or motherboard. Thus, when signal runs are greater than the *critical line length* [see Equation (9-1) for the definition of *critical line length*], signal quality will degrade as a result of transmission-line effects, and additional time, beyond the one-way propagation delay of the lines, will be required for signals to settle.

As unterminated signal lines increase in length beyond *the critical length*, signal quality progressively degrades and takes longer to stabilize. Depending upon where along a line a signal is observed, transmission-line effects will be apparent on lines longer than *the critical line length* under most conditions, even when lines are properly terminated. When signal lines are not significantly longer than *the critical length,* transmission-line effects may not be severe enough to impact normal operation. When TTL components are used, the possible detrimental response of lines may be controlled by the current-limiting resistors used in the pull-up section of TTL totem-pole output stages and the input clamp diodes that are built into the input-output circuits of most TTL devices (see Chapter 3). When CMOS components are used, the situation is less clear. There is little uniformity in the input-output circuitry used or of the response of CMOS input-output circuits. Thus, when CMOS logic components are used and the interconnecting lines are longer than *the critical length*, additional time beyond one-way line propagation times must be allowed for signals to settle unless the lines are properly terminated. Proper termination of a line consists of matching the impedance of the source and the line, or the load and the line, or both. It is generally impractical to terminate all signals in a digital system since terminations require additional components and may dissipate additional power (depending upon the method of termination used).

11-5-2 Transmission lines

A representation of a digital signal transmission line that has one source and one load is shown in Figure 11-7.[15] The lossless ideal

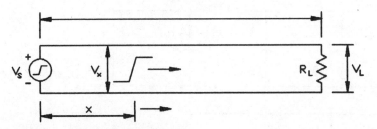

Figure 11-7 Typical transmission line.

equivalent circuit for a section of such a transmission line (i.e., the one shown in Figure 11-7) is shown in Figure 11-8.[16] At a sufficiently high frequency, which is exceeded by the frequency components in the switching edges of advanced Schottky and advanced CMOS devices, the inductive and capacitive effects cancel and a transmission line appears to the source and load as a pure resistance.[17] The apparent resistance of a transmission line is called the characteristic impedance Z_o of the line.

The characteristic impedance Z_o of an ideal transmission line is defined as[13,18,19]

$$Z_o = \sqrt{\frac{L}{C}} \tag{11-8}$$

and the propagation delay t_{pd} as

$$t_{pd} = \sqrt{LC} \tag{11-9}$$

where L and C are as shown in Figure 11-8.

From a practical standpoint Equations (11-8) and (11-9) are of limited value since L and C are difficult to determine. However, most texts that deal with transmission lines show empirically derived equations for determining the characteristic impedance and propagation delay time for most of the common physical interconnection structures.

For wire-wrap or stitch-weld boards, the basic physical interconnection structure is a wire over a ground or voltage (V_{cc}) plane (voltage planes serve as ac references) as shown in Figure 11-9.

For such a configuration, Z_o can be calculated from[4,11]

$$Z_o = \frac{60}{\sqrt{\epsilon_r(\text{eff})}} \ln\left(\frac{4h}{d}\right) \tag{11-10}$$

where d is the wire diameter, h is the distance from the ground to the

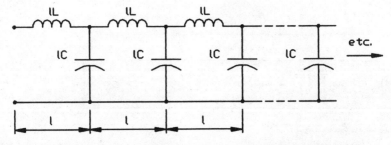

Figure 11-8   The equivalent circuit for a section of an ideal transmission line.

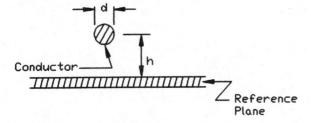

Figure 11-9 Wire over a reference plane.

center of the wire, and t_{pd} can be calculated from

$$t_{pd} = 1.017\sqrt{\epsilon_r(\text{eff})} \qquad (11\text{-}11)$$

where the constant 1.017 is the reciprocal of the velocity of light in free space.[13,19] In both Equations (11-10) and (11-11), ϵ_r is the effective dielectric constant of the material in the separation between the reference plane and the wire. For a wire in air the dielectric constant is 1, but the case of a wire in air is of little interest. For an actual wire-wrap or stitch-weld board, the separation consists of air, the insulation on the wire, and other conductors. Thus, the effective dielectric constant is difficult to determine, and Z_o and t_{pd} calculations are quite complex. Actual Z_o measurements of wire-wrap and stitch-weld boards show Z_o to be in the range of 80 to 150 Ω and t_{pd} to be on the order of 1.5 ns/ft. The value of Z_o is dependent on the distance between the wires and the reference plane. If the wires are not kept close to the reference plane, Z_o values higher than 150 Ω are possible.

For pc board interconnections there are two basic physical arrangements of conductors possible relative to the reference planes: a conductor above a reference plane (referred to as microstrip) or a conductor enclosed between two reference planes (referred to as stripline).

A microstrip consists of a thin rectangular conductor above a ground or reference plane, as shown in Figure 11-10a and b. A microstrip conductor can be on the surface of an insulating dielectric, as shown in Figure 11-10a, or buried in a dielectric, as shown in Figure 11-10b. Configuration a corresponds to conductors on uncovered surface layers of pc boards and configuration b to conductors on internal layers of multilayer pc boards (which are not enclosed by reference planes). Figure 11-10b is the more common situation since most present-day high-performance systems are built with multilayer pc boards.

The equation for Z_o (in ohms) for a microstrip transmission line on the surface of a dielectric is[20,21]

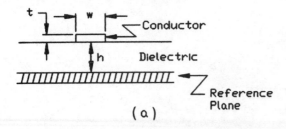

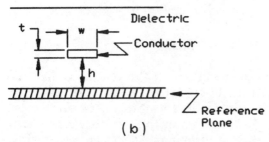

Figure 11-10 (a) Microstrip on the surface of a dielectric. (b) Microstrip buried in a dielectric.

$$Z_o = \frac{60}{\sqrt{0.475\epsilon_r + 0.67}} \ln\left(\frac{5.98h}{0.8w + t}\right) \qquad (11\text{-}12)$$

and t_{pd} for a surface microstrip (in nanoseconds per foot) is

$$t_{pd} = 1.017\sqrt{0.475\epsilon_r + 0.67} \qquad (11\text{-}13)$$

where h, t, and w are in inches and are shown in **Figure 11-9 and ϵ_r is the dielectric constant of the material between the conductor and the reference plane.**

The equation for Z_o (in ohms) for a buried microstrip transmission line is

$$Z_o = \frac{60}{\sqrt{\epsilon_r}} \ln\left(\frac{5.98h}{0.8w + t}\right) \qquad (11\text{-}14)$$

and t_{pd} for a buried microstrip (in nanoseconds per foot) is

$$t_{pd} = 1.017\sqrt{\epsilon_r} \qquad (11\text{-}15)$$

(The variables are as defined above.)

A stripline transmission line consists of a thin rectangular conductor with a reference plane above and below the conductor, as shown in Figure 11-11.

The equation for Z_o (in ohms) for a stripline conductor is[20,21]

$$Z_o = \frac{60}{\sqrt{\epsilon_r}} \ln \left[\frac{4b}{0.67\pi(0.8w + t)} \right] \qquad (11\text{-}16)$$

$$t_{pd} = 1.017\sqrt{\epsilon_r} \qquad (11\text{-}17)$$

where b, h, t, and w are in inches and are shown in Figure 11-11.

The dielectric constant ϵ_r for epoxy glass pc board material is approximately 5. For interconnection layers on most present-day multilayer pc boards, 1-oz copper plating is used. The thickness t in Equations (11-14) and (11-16) for 1-oz copper is approximately 0.0015 in (t for 2-oz copper is 0.003 in). References 13, 16, 21, and 22 include figures that show the characteristic impedance of epoxy glass pc boards for various conductor sizes and spacings relative to a reference plane (or planes). Typical pc board conductor-reference plane spacing results in a Z_o that is in the range of 30 to 70 Ω and a t_{pd} that is in the range of 1.8 to 2.2 ns/ft.

In addition to wires over grounds and microstrip and stripline conductor configurations, digital designers must often deal with coaxial cables and twisted-pair interconnections. Cable parameters, such as Z_o and t_{pd}, are supplied in cable manufacturers' catalogs (which should be consulted when such interconnections are used). Coaxial cables are available with various Z_o values; typical values are in the 50- to 90-Ω range.[23] Twisted-pair lines have Z_o in the neighborhood of 120 Ω, and shielded twisted-pair lines are in the 70- to 120-Ω range.[23]

11-5-3 Loaded transmission lines

The effective characteristic impedance Z_o' and effective propagation time t_{pd}' of a line is a function of the dynamic load C_{LOAD}. The expres-

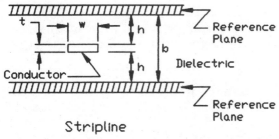

Stripline

Figure 11-11 Stripline.

sion for the effective characteristic impedance of a loaded line is[4]

$$Z_o' = \frac{Z_o}{\sqrt{1 + (C_{LOAD}/C_{LINE})}}$$ (11-18)

and for the effective propagation delay is

$$t_{pd}' = t_{pd}\sqrt{1 + \frac{C_{LOAD}}{C_{LINE}}}$$ (11-19)

where C_{LOAD} is the total lumped capacitance (inputs and outputs) of each device connected to the line and C_{LINE} is the total line capacitance. Determining C_{LOAD} is straightforward; C_{LOAD} is the sum of the input, or output capacitance, as the case may be, of all of the devices connected to the line. Determining C_{LINE} requires that the per unit length line capacitance C be known ($C_{LINE} = C \times$ line length). However, deriving C (per unit length) is not a straightforward process. One approach is to use the basic equation for Z_o, i.e., Equation (11-8),

$$Z_o = \sqrt{\frac{L}{C}}$$

and rearrange it to

$$C = \frac{L}{Z_o^2}$$ (11-20)

to calculate C per unit length of line. The characteristic impedance Z_o used in Equation (11-20) can be calculated from equations such as (11-12), (11-14), and (11-16) or may be available from the manufacturer's specifications, but L in Equation (11-20) is not readily available and is difficult to calculate.

The most practical means of determining the capacitance per unit length of a line C is to measure the propagation delay of an actual unloaded pc board track of the configuration of interest and use the measured time t_{pd}(measured) and Z_o(calculated)—from Equations (11-12), (11-14), or (11-16)—in Equation (11-23), which is derived from Equations (11-21) and (11-22), which are, in turn, specific cases of Equations (11-8) and (11-9).

$$t_{pd}(\text{measured}) = \sqrt{LC}$$ (11-21)

$$Z_o(\text{calculated}) = \sqrt{\frac{L}{C}}$$ (11-22)

$$C = \frac{t_{pd}(\text{measured})}{Z_o(\text{calculated})} \qquad (11\text{-}23)$$

and the total capacitance of the line C_{LINE} is

$$C_{\text{LINE}} = C \times \text{line length} \qquad (11\text{-}24)$$

where the line length is in the same length units as those used with C.

It is important to note that lines with a number of loads will have effective Z_o' that is much less than the unloaded Z_o and that the effective t_{pd}' will be much greater. It is not uncommon for the effective Z_o' of a loaded pc track to be as low as 20 Ω and the actual propagation delay as large as 3.5 ns/ft.

11-5-4 Transmission-line response

A long transmission line, with a one-way delay greater than the signal rise time, with one source and one load, responds to signal transitions approximately as follows:

1. A signal transition is initiated at the source.
2. The signal travels along the transmission line with a delay time that depends on the dielectric constant of the surrounding material.
3. The signal arrives as the first incident wave at the load after a delay that is a function of the line length and the dielectric constant of the material adjacent to the line.
4. Part of the energy of the first incident wave is absorbed by the load in establishing the initial signal level at the load.
5. The part of the energy of the first incident wave that is not absorbed as a result of a mismatch between the line and the load impedance is reflected back toward the source.
6. When the reflection of the first incident wave arrives at the source, part of it may be absorbed by the source in establishing a new signal level; any mismatch is reflected back toward the load.
7. When the second incident wave arrives at the load, part of the energy is absorbed in establishing a new signal level; any mismatch is reflected back to the source as in item 5 above.
8. The cycles of reflections between load and source are repeated with part of the energy absorbed in each cycle; after some time, depending upon the characteristics of the source, line, and load, the reflected transient levels are so small that they are negligible.

Reflections.   The amount of a signal that is reflected when an incident wave arrives at a load is determined by the mismatch between the impedance of the line and the load (or source). If a transmission line is terminated with an impedance that is equal to the characteristic impedance Z_o of the line, there will be no reflection from the end of the line, and the only signal appearing on the line will be the incident wave. If some other value of termination is used, a portion of the incident wave will be reflected and the signal appearing on the line will be the sum of the incident and reflected waves. The magnitude and polarity of the reflection from a load or a source is quantitatively described by the reflection coefficient ρ.

The equation for the reflection coefficient ρ_L for the load end of a line is[24,25]

$$\rho_L = \frac{R_L - Z_o}{R_L + Z_o} \tag{11-25}$$

and the reflection coefficient ρ_S for the source end of the line is

$$\rho_S = \frac{R_S - Z_o}{R_S + Z_o} \tag{11-26}$$

If either end of the line is exactly matched to Z_o, that is, $R_S = Z_o$ or $R_L = Z_o$, the reflection coefficient is 0 and the incident wave is completely absorbed and no reflection occurs.

The action of a transmission line with nonzero reflection coefficients is shown diagrammatically in Figure 11-12 (the dimensions x and l are shown in Figure 11-7).

There are a couple of cases of ρ that are of special interest. They are:

1. $\rho = +1$ when $R_L = \infty$ or $R_S = \infty$ (an open-circuited line).

   When $\rho = +1$ the signal doubles when the incident wave arrives at the end of the line.

2. $\rho = -1$ when $R_L = 0$ or $R_S = 0$ (a short-circuited line).

   When $\rho = -1$ the incident wave reverses its polarity and subtracts an amount equal to the incident wave from the existing voltage at the load (or source) and the new voltage level is reflected back toward the other end of the line. The energy in an incident wave is *not* absorbed by a short.

Case 1, above, is the typical situation for TTL or CMOS logic circuits; inputs are sensitive to voltage (not power) and require only a small proportion of the energy in the incident wave arriving from the transmission line to maintain steady-state operation (i.e., the load impedance is very large with respect to the line impedance). Therefore,

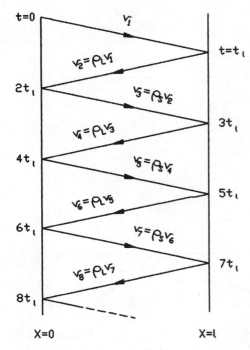

t=0

v_i

$t=t_1$

$v_2 = \rho_L v_1$

$2t_1$

$v_3 = \rho_s v_2$

$3t_1$

$v_4 = \rho_L v_3$

$4t_1$

$v_5 = \rho_s v_4$

$5t_1$

$v_6 = \rho_L v_5$

$6t_1$

$v_7 = \rho_s v_6$

$7t_1$

$v_8 = \rho_L v_7$

$8t_1$

X=0   X=l

Figure 11-12 Reflection diagram.

the voltage at the load device is increased (up to double) compared to the incident wave and a significant portion of the energy is reflected back toward the source.

Case 2, above, is approximated by typical CMOS *high* and *low* outputs and TTL *low* outputs; their effective output impedance is not zero, but it is much less than normal line impedance. Thus, *low* TTL outputs, or CMOS *high* or *low* outputs, have negative reflection coefficients. Energy is reflected with a reversal of voltage polarity. A negative reflection coefficient at the source end of a line will cause overshoot that may have occurred at the load end of the line to be converted to undershoot on alternate reflection cycles. The undershoot, when it arrives at the load, may cause the signal level to transition into the threshold region of the receiving device, upsetting the logic sense of the signal. Under such conditions, which are typical of advanced TTL or advanced CMOS devices, a number of round trips may be required for the excess energy in the signal to be absorbed and for the signal to achieve a steady-state value. Figure 11-13 illustrates the response of a typical line with one source and one load when driven and received with devices with no input clamp diodes (for example, *high* TTL-level signals usually are of insufficient magnitude to go into the clamping region).[1] Figure 11-13 is drawn for a load reflection coefficient of +1 and a source reflection coefficient of −0.5; both

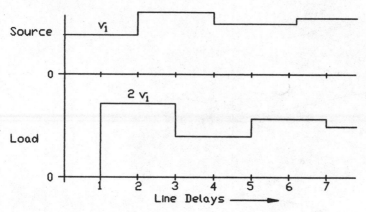

Figure 11-13   Typical response of a long line.

values are typical of those encountered in most TTL or CMOS applications.

In the case illustrated in Figure 11-13, the waveform at the load end does not undershoot (on the second incident wave) below V_1, but in many actual cases the signal will undershoot back into the threshold region of the receiving device and will remain there until the signal travels back to the source and back to the load. Thus, for most lines that are not terminated, five line delays must be allowed for signals to settle. After five line delays, most of the energy in a wave front will have been absorbed and the undershoot will have effectively dampened out in most cases. However, it is difficult to generalize a very complex phenomena. Critical situations must be carefully analyzed.

A rule of thumb that will serve for most timing analysis is: *Allow five line delays for unterminated lines to settle.*

When performing worst-case timing analysis, designers should allow for any additional signal length that might be added to signal paths during system tests or troubleshooting. One example is the added circuit board to motherboard signal length that occurs when extender boards are used to extend circuit boards for troubleshooting.

11-5-5   Transmission-line termination

There are two basic ways of terminating transmission lines—series (source) termination or load termination. Either method can be used to achieve a stable signal at the far (load) end of a line after one line delay. However, the response at the source and at intermediate points is different for the two methods of termination. Series termination requires at least two line delays for signals to stabilize at the source end of a line; load-terminated lines can reach steady-state conditions at the source, at all points along the line, and at the load after one line delay.

Load termination. A line is defined as load terminated when the load at the end of the line is matched to the impedance of the line. Figure 11-14a, b, c, and d shows various means of load termination.[4,26] The same dynamic results are achieved with each of the configurations.

When a line is load terminated with a matching impedance (R_L of the load = Z_o of the line), regardless of the dc circuit configuration of

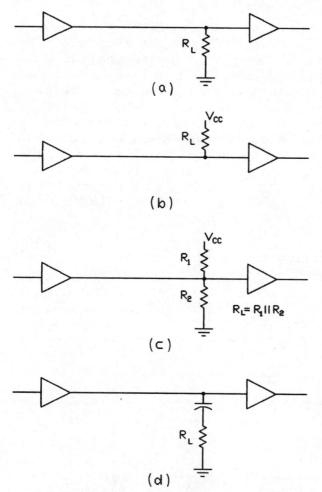

(a)

(b)

(c)

$R_L = R_1 \| R_2$

(d)

Figure 11-14 Load termination. (a) Termination to ground; (b) termination to V_{cc}; (c) split termination; (d) ac termination.

the terminating network, the signal at the source, intermediate points, and at the load should appear as a well-behaved signal as shown in Figure 11-15. In such a case, the signal is launched into the line and at some later time (equal to the propagation time of the line t_{LINE}) arrives at the load. The energy in the wave front is absorbed, and a steady-state level is established upon the arrival of the incident wave at the load (see Figure 11-15). In Figure 11-14 it is implied that the receivers have infinite impedance, and for all practical purposes, that is the case for TTL and CMOS device inputs, as long as the signal levels remain between ground and V_{cc}. Thus, the value of the load termination does not need to be adjusted to compensate for receiver input impedance.

Source termination.   Source termination, which is often referred to as series termination, consists of matching the source impedance to the line impedance, as shown in Figure 11-16. The waveforms at the source, intermediate points, and the load for a source-terminated circuit that is exactly matched to the line and that has an infinite load are shown in Figure 11-17.

Series termination works best where there is one source and one load. The waveform at the load end of a series-terminated line is well behaved, but at the source and at intermediate points near the source, the leading edge of the waveform will step up rather than make a smooth transition (see Figure 11-17). The initial step (or steps) occurs as a result of the divider formed by R_S and Z_o of the line (see Figure 11-18). The amplitude of the initial step is

$$V_{t=0}(\text{at the source}) = \left(\frac{Z_o}{R_S + Z_o}\right)(V_{cc}) \qquad (11\text{-}27)$$

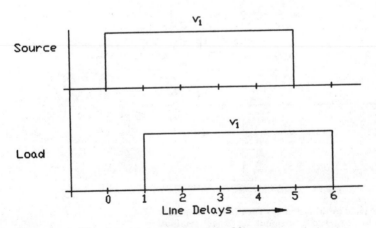

Figure 11-15   Waveforms in a load-terminated line.

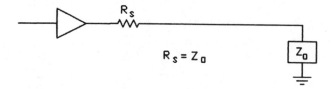

Figure 11-16 Source termination.

When R_S is matched to Z_o, the amplitude of the initial step is

$$V_{t=0}(\text{at the source}) = \tfrac{1}{2}V_{cc} \qquad \text{when} \qquad R_S = Z_o \qquad (11\text{-}28)$$

However, if the Z_o of a line is very low relative to R_S, multiple energy injection and reflection cycles and steps may be necessary for the line to reach its final value.

Actual TTL output circuits are not quite as simple as the circuit in Figure 11-18; *high*-level and *low*-level output impedances of TTL devices are significantly different. For standard (i.e., not drivers) advanced Schottky components, *high*-level output impedance is near 40 Ω and *low*-level output impedance is near 10 Ω. Thus, even with the addition of an external series resistor, an exact impedance match to a line cannot be achieved for both *high*- and *low*-going TTL signals. Hence, series termination of TTL circuits is by necessity a compromise. Since *low* TTL levels have the least noise margin and since all TTL devices have higher drive in the *low* state, which increases the

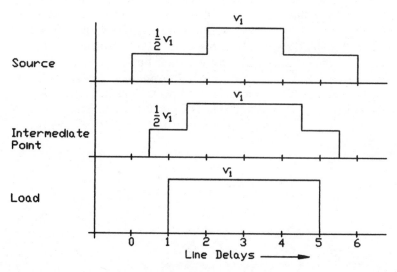

Figure 11-17 Waveforms in a source-terminated line.

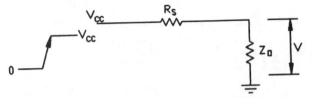

Figure 11-18 Output voltage divider formed by series termination and the line impedance.

possibility of ringing, it is best to match more exactly the *low*-state output impedance.

The *high* and *low* output impedance of CMOS devices tends to be more evenly matched than that of TTL devices, but CMOS output impedance changes on the order of two-to-one from nominal in both the positive and negative direction with processing, temperature, and voltage. Thus, it is impossible to closely match CMOS outputs to line impedances.

The real world responses of TTL or CMOS signals, whether source terminated or not, do not consist of nice single-level square-cornered transitions at sources or at intermediate points in lines. Actual responses tend to look like the waveform shown in Figure 11-19.

Real TTL and CMOS devices have some finite amount of output impedance. Hence, signal waveforms similar to the one shown in Figure 11-19 are common throughout TTL and CMOS systems; in most cases they do not cause a problem.[24,27,28] Such signal responses are of no concern in synchronous systems on data and control lines. In a few remote cases, a waveform with a step can cause problems at a secondary level. If a step happened to be near the threshold level of a receiver, the receiver input circuitry may oscillate during the time the signal dwells in the threshold region. In some cases, the oscillation may generate enough noise to upset nearby circuits, but in most cases, the time that signals dwell in the intermediate region is relatively short, and as a result no ill effects are caused. However, such signals cannot be used to clock devices; multiple clocking may occur. Since signals with steps are inherent near the source of a TTL or CMOS driven line, high-speed

Figure 11-19 Actual signal waveform near the source end of a line with finite source impedance.

TTL and CMOS clock lines must be arranged so that the clocked devices are near the ends of the lines (see Chapter 9 for clock distribution guidelines) or the impedance of the lines must be high enough that the level of the step is well above the threshold of the clocked devices.

Minimum line impedance requirements in TTL systems are determined by device *high* output impedance. Since advanced Schottky devices typically have 45-Ω output-stage pull-up resistors, clock lines, or other lines where wave shape is important need to have an effective line impedance on the same order (40 Ω is a practical lower limit) to ensure that the initial *high*-going signal transition crosses the switching threshold. If the interconnection system minimum line impedance meets the needs of *high*-going signals, it will be more than adequate for *high*-to-*low* TTL transitions since *low* TTL device output impedance tends to be near 10 Ω.

Forty ohms is also near the minimum usable line impedance for CMOS signals that must switch on the first incident wave. Advanced CMOS input thresholds are at 30 and 70 percent of V_{cc}. Thus, both *high* and *low* initial signal transitions must equal or exceed two-thirds of V_{cc}. Under nominal +25°C conditions the output impedance of advanced CMOS devices is near 10 Ω. Worst-case output impedance over the extremes of military operating conditions is near 20 Ω. Thus, initial signal transitions on lines with 40 Ω or greater impedance will equal or exceed two-thirds of V_{cc}.

11-6   Signal Quality Summary

1. Static (dc) loading should not exceed 70 percent of the manufacturer's rated drive capability when system operation is over the full military temperature range, or 80 percent when system operation is over the commercial temperature range (unless more conservative drive derating is required by the customer).

2. Signals that are highly susceptible to crosstalk should be run at right angles to one another or be isolated by reference planes (when multilayer pc boards are used). For example, memory address lines should run at right angles to memory data lines to minimize coupling between address and data lines.

3. Signals that are highly sensitive to crosstalk, such as clocks, should be isolated from other signals (including other clocks) by a distance of at least twice the distance to the closest reference plane.

4. When signals are sent between circuit boards, consideration must be given to possible signal degradation due to transmission-line

effects since most board-to-board interconnections will be long. Board-to-board communication must also consider possible added signal length due to extender boards, since digital systems must function with boards on extender boards when troubleshooting is necessary. Unterminated board-to-board and unit-to-unit signal interconnections must allow five line delays for ringing and reflections to subside.

5. Signal lines with wave-shape requirements (such as clocks) must be terminated.

   Series termination is best for most single-source to single-load applications.

   Load termination is best for single-source to multiple-load applications.

   Terminations that result in slightly underdampened lines provide optimum response.

6. Interconnection line impedance should be greater than 40 Ω (unloaded) so that TTL or CMOS signals will not be excessively degraded.

7. Signals routed to test points should be buffered to prevent test equipment from interfering with normal system operation.

8. Signals that are used internally and externally to a board or unit should have external lines buffered to isolate internal signals from transmission-line reflections or other external disturbances such as shorts.

9. Input signals to boards, racks, or systems should be buffered at one place so as to minimize the loading of the source.

10. Low duty cycle control lines with TTL levels should be arranged so that they are at a *high* level most of the time since TTL devices have maximum noise margin in the *high* state.

References

1. Nguyen-huu, Anh: "An Analysis of the Ringing Phenomenon in Digital Systems," *Computer Design,* July 1971, pp. 39–45.
2. Davidson, Malcolm: "Understanding the High Speed Digital Logic Signal," *Computer Design,* November 1982, pp. 79–82.
3. Royle, David: "Transmission Lines and Interconnections," Part One, *EDN,* June, 23, 1988, pp. 131–136.
4. *FAST Applications Handbook 1987,* National Semiconductor Corp., South Portland, Me., 1988.
5. Southard, Robert K.: "High-Speed Signal Pathways from Board to Board," in 1981 WESCON Rec., Session 18, September 1981, Paper No. 2.
6. Gheewala, Tushar, and David MacMillan: "High-Speed GaAs Logic Systems Require Special Packaging," *EDN,* May 17, 1984, pp. 8–14.
7. Kozuch, John J.: "A High Speed Approach to Controlled Impedance Packaging," Multiwire Div. Kollmorgen Corp., January 1987.

8. Mohr, Richard J.: "Interference Coupling—Attack it Early," *EDN*, July 1, 1969, pp. 33–41.
9. DeFalco, John A.: "Reflection and Crosstalk in Logic Circuit Interconnections," *IEEE Spectrum*, July 1970, pp. 44–50.
10. DeFalco, John A.: "Predicting Crosstalk in Digital Systems," *Computer Design*, June 1973, pp. 69–75.
11. Springfield, William K.: "Designing Transmission Lines into Multilayer Circuit Boards," *Electronics*, November 1, 1965, pp. 90–96.
12. Harper, Charles A.: *Handbook of Electronic Packaging*, McGraw-Hill, New York, 1969.
13. Balph, Thomas: "Implementing High Speed Logic On Printed Circuit Boards," in 1981 WESCON Rec., Session 18, September 1981, Paper No. 1.
14. Royle, David: "Transmission Lines and Interconnections," Part Three, *EDN*, June 23, 1988, pp. 155–160.
15. Millman, Jacob, and Herbert Taub: *Pulse, Digital and Switching Waveforms*, McGraw-Hill, New York, 1965.
16. *F100K ECL Logic Databook and Design Guide*, 1989 ed., National Semiconductor Corp., Santa Clara, Calif., 1989.
17. Arvanitakis, N. C., and J. J. Zara: "Design Considerations of Printed Circuit Transmission Lines for High Performance Circuits," in 1981 WESCON Rec., Session 18, September 1981, Paper No. 4.
18. Matick, Richard E.: *Transmission Lines for Digital and Communications Networks*, McGraw-Hill, New York, 1969.
19. Kaupp, H. R.: "Characteristics of Microstrip Transmission Lines," *IEEE Trans. Electronic Computers*, Vol. EC-16, No. 2, April 1967, pp. 185–193.
20. *F100K ECL User's Handbook*, Fairchild Semiconductor Corp., Puyallup, Wa., 1986.
21. Blood, William R., Jr.: *MECL System Design Handbook*, 4th ed., Motorola Semiconductor Products Inc., Phoenix, Az., 1988.
22. Harper, Charles A.: *Handbook of Components for Electronics*, McGraw-Hill, New York, 1977.
23. Crouch, Ronald A.: "Choose Cable with Care to Optimize System Design," *EDN*, November 5, 1978, pp. 113–116.
24. Saenz, R. G., and E. M. Fulcher: "An Approach to Logic Circuit Noise Problems in Computer Design," *Computer Design*, April 1969, pp. 84–91.
25. DeClue, Joseph L.: "Wiring for High-Speed Circuits," *Electronic Design*, No. 11, May 24, 1976, pp. 84–86.
26. Royle, David: "Transmission Lines and Interconnections," Part Two, *EDN*, June 23, 1988, pp. 143–148.
27. Heniford, William: "Muffling Noise in TTL," *The Electronic Engineer*, July 1969, pp. 63–69.
28. Burton, Edward A.: "Transmission-Line Methods Aid Memory-Board Design," *Electronic Design*, December 8, 1988, pp. 87–97.

Bibliography

Blood, William R., Jr.: *MECL System Design Handbook*, 4th ed., Motorola Semiconductor Products Inc., Phoenix, Az., 1988.
Harper, Charles A.: *Handbook of Wiring, Cabling and Interconnections for Electronics*, McGraw-Hill, New York, 1972.
Matick, Richard E.: *Transmission Lines for Digital and Communications Networks*, McGraw-Hill, New York, 1969.
Porter, Jack: "Compute Coupled-Microstrip Line," *Electronic Design*, No. 1, January 4, 1977, pp. 116–118.
Skilling, Hugh H.: *Electric Transmission Lines*, McGraw-Hill, New York, 1951.

System Timing

High-speed high-performance digital systems require a careful timing analysis of all critical signal paths to establish the maximum usable system clock frequency or to establish that the critical path delays are compatible with a predetermined clock frequency. Very often unrealizable operating speeds are forecast for advanced Schottky and advanced CMOS systems because of improper allowances for worst-case component delays and little or no allowance for signal wiring propagation delays and transmission-line effects. For a realistic estimate of the operating speed of a system, the worst-case propagation delays of each component in each signal path, plus the physical propagation delays and settling time needed for each interconnection network must be determined and factored into the projected operating speed of each design. Signal interconnection and settling time delays are a large part of most signal propagation delay paths when system operating speeds approach 20 MHz or greater.

At present, most system- and board-level timing analyses must be done by hand. Computer-aided design (CAD) tools that combine logic device timing and pc boards or backpanel interconnection system transmission-line response characteristics are not available to most designers. Thus, high-speed systems must be structured so that a thorough timing analysis is a manageable task. Synchronous design is one means of simplifying system timing analysis. In synchronous designs, data tend to flow in an orderly manner from one clocked device to another with signal path propagation requirements clearly defined by the system clock period. When a synchronous architecture is used, a system is organized into manageable and more readily understood data paths. In contrast, in asynchronous systems, data tend to flow through long irregular paths that are dependent on many conditions with the consequence that a proper timing analysis is usually difficult or impossible to achieve.

12-1 Device Delays

Manufacturers' device specification sheets or data books all contain some timing information. Unfortunately, there is little uniformity with respect to the load and test conditions used to define timing parameters. Variations in load capacitance and resistance used for specified timing makes interpreting and comparing timing specifications difficult. To further complicate matters, worst-case timing (over temperature, supply voltage, etc.) is not provided by most manufacturers for the older logic families (LS, S, etc.).

12-1-1 Device timing adjustments for temperature and voltage levels

Most manufacturers of advanced Schottky or advanced CMOS devices specify worst-case propagation delays, and setup and hold times over the allowed operating conditions (i.e., supply voltage, temperature, load, etc.). Thus, for most operating conditions, the manufacturer's maximum and minimum specified timing can be used without derating (except for load capacitance in excess of 50 pF).

Since worst-case timing is not specified for most devices in the older logic families, some derating criteria must be applied for worst-case conditions. For most older devices, only maximum propagation times (minimum for setup and hold times) are given for ideal operating conditions, i.e., $+25°C$, V_{cc} at $+5$ V, and 15-pF loads. If a system must operate under conditions other than ideal, the timing must be derated to allow for timing degradation that will occur as a result of worse-than-nominal conditions.[1]

When only $+25°C$ and $+5$-V V_{cc} *maximum* (or minimum) timing values are given on data sheets, those values should be multiplied by 1.5 to establish the derated timing values for parts used in systems that must operate over the full military temperature range and supply-voltage limits (i.e., -55 to $+125°C$ and 4.5- to 5.5-V V_{cc}). When only *typical* $+25°C$ timing values are given, a derating factor of 2 should be used to compensate for device degradation over worst-case military operating conditions.

For less severe applications, such as 0 to $+70°C$ applications, a derating factor of 1.25 is typically applied to *maximum* ideal timing parameters and a factor of 1.5 to *typical* ideal values.

Summary of Device Worst-Case Propagation Delay Specifications and Adjustments Needed for Full Temperature Range Operation

1. Data sheets provide worst-case timing parameters for the AC, AS, HC, ALS, and FAST logic families. Both the worst-case military (-55 to $+125°C$) and commercial (0 to $+70°C$) values are specified on most data sheets.

2. Most data sheets do not provide worst-case timing parameters for the LS and S logic families. A derating factor of 1.5 must be applied to the $+25°C$ *maximum* values for full military temperature range operation and a factor of 1.25 for commercial temperature range operation. When only *typical* $+25°C$ timing parameters are provided, a derating factor of 2 is needed for full military temperature range operation and a factor of 1.5 for commercial temperature range operation.

12-1-2 Device timing adjustments for actual load conditions

In addition to basic device derating for temperature and supply-voltage levels, a timing analysis must take into consideration the actual loading if the loading is different (which it will usually be) from the value at which the timing is specified. The dynamic response of most devices in the advanced logic families is specified with 50-pF loads. In contrast, most of the older logic families are specified with 15-pF loads. In most cases, though, 50 pF is a much more realistic load, and in many situations actual loads will exceed 50 pF, particularly in many bus driving applications. In those cases where loads exceed 50 pF (or 15 pF for LS and S), some adjustment must be made to the specified output response time of the driving device. In some cases, device manufacturers provide generic graphs that show the typical change in output response versus load capacitance. Figure 12-1 is a compilation of such data for the S, AC, HC, LS, ALS, and FAST logic families.[2–5]

Note that the timing adjustment graphs in Figure 12-1 are for room temperature and nominal supply-voltage conditions. Worst-case response versus load capacitance is seldom provided by device manufacturers. A reasonable estimate of worst-case values is provided by multiplying typical values by a factor of 2. The factor of 2 is based on possible device characteristic changes due to process variations and operation at the limits of the temperature and supply voltage range.

The following maximum delay versus load capacitance adjustment factors were derived by applying a factor of 2 to the typical delay versus load data shown in Figure 12-1.

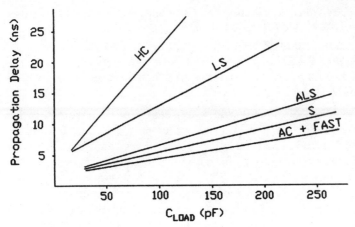

Figure 12-1   Output response versus load capacitance.

200 ps/pF × ΔC for HC

150 ps/pF × ΔC for LS

80 ps/pF × ΔC for ALS

50 ps/pF × ΔC for S, AC, FAST, and LS and ALS drivers

where ΔC is the actual load minus the specified load. The above delay versus load factors can be used to adjust the manufacturers' maximum timing values for load capacitance in excess of that at which device timing is specified or to adjust minimum timing values when the load is less than the specified load.

When delay versus load capacitance is not provided, an estimate of the increased delay due to larger-than-specified load capacitance can be determined from Equation (11-1).

$$\frac{dv}{dt} = \frac{I_{SC}}{C}$$

where I_{SC} is the output short circuit current of the output driving device and C is the increase in load capacitance beyond the specified load value. For TTL devices or CMOS devices with TTL output levels, the value of I_{SC} is different for *high* and *low* outputs. In most cases, *low*-to-*high* transition times are longer than those of *high*-to-*low* transitions. Most TTL devices have lower *high*-level I_{SC} than *low*-level I_{SC} because of the current-limiting resistors used in TTL output pull-up circuits. Most CMOS devices that have TTL output levels use *n*-channel source followers for output pull-ups. Such circuits tend to have less *high* output drive than conventional switched CMOS outputs.

Line capacitance.   The total capacitance of a line consists of the input or output capacitance of each device connected to the line plus the line capacitance.

Device input or output capacitance is usually included on data sheets. If it is not, capacitance loading due to standard SSI or MSI device connections can be estimated using the following values:

- 5 pF for inputs
- 7 pF for outputs
- 15 pF for bidirectional ports

However, device specifications should be consulted when they are available to ensure that the proper values are used. Programmable devices, such as PALs, should be viewed with caution; PAL input capacitance may be as large as 20 pF.

Typical interconnection (wiring) capacitances are as follows:

- 3 to 5 pF/in of pc board track
- 1 to 2 pF/in for welded-wire or wire-wrap wire

Actual interconnection capacitance should be used in timing calculations when available. However, in many applications, the actual capacitance of a given signal interconnection is difficult to determine because of the great variations in conductor and ground reference spacing. Wire spacing relative to reference plans or other conductors on stitch-weld or wire-wrap boards varies a great deal. Even where conductor topology is known, it is difficult to determine exact line capacitance because of the interaction of nearby conductors. Thus, most line capacitance calculations are made assuming only an isolated conductor and a nearby reference plane.

For most applications, lumped loading due to interconnection (wiring) capacitance can be estimated using the following values:

- 4 pF/in of pc board track
- 1.5 pF/in for welded-wire or wire-wrap wire

Guidance for calculating more exact interconnection capacitance values can be found in a number of the transmission-line references listed at the end of Chapter 11.

In most practical circuit applications, LS- and S-specified 15-pF load capacitance is exceeded. Thus, when LS and S are used, manufacturer-supplied timing values must be derated for additional loads. When the advanced logic families are used, the 50-pF rating is seldom exceeded except for bus driving applications. However, most busses are long line, and thus are not lumped loads and should not be treated as such.

12-1-3   Caution

The parameter F_{MAX}, which is listed as the maximum toggle rate or maximum clock rate on data sheets for clocked devices such as counters, flip-flops, and shift registers, should never be used as an indication of the useful speed of a device.[1] It is a measure of what might be achieved with an individual part under ideal conditions with no restrictions on input pulse widths or load conditions.[1] Since most digital systems must operate under conditions that vary greatly from ideal and since devices must communicate with other devices to serve a useful purpose, F_{MAX} is of little use for actual system timing. Actual signal path propagation times must be used to determine the maximum operating speed of systems, not F_{MAX}.

12-2   Circuit Board Interconnection Delays

Interconnection delays consist of conductor propagation delay and signal settling time delay. Signal propagation delay is the time required for a signal to traverse a physical conductor. Signal settling time is the time required for a signal to settle to a proper logic level. Signal settling time is most often of concern when signal lines are long, i.e., in a transmission-line environment.

Calculating conductor physical delay is straightforward. Conductor delay is equal to the per unit length propagation delay t_{pd} of electrical energy in the interconnection media multiplied by the length of the interconnection. For pc boards t_{pd} is approximately 2 ns/ft, and for welded-wire or wire-wrap boards t_{pd} is approximately 1.5 ns/ft. For extremely time critical circuits, very long lines (several feet), or very heavily loaded lines, more exact propagation delay times should be calculated (see Section 11-5 and the various transmission-line references listed at the end of Chapter 11). However, for most applications the per unit length signal propagation values given above are sufficient.

Calculating interconnection delays due to load and transmission-line effects are not straightforward, yet both effects may add significant time to signal delays. Transmission-line effects become of concern in high-speed TTL and CMOS systems when line lengths exceed the *critical line length* (see Section 11-5). The common definition of *critical line length* is given by Equation (11-5):

$$\text{Critical line length} = \frac{1}{2}\frac{t_r}{t_{pd}'}$$

where t_r is the rise time of the driving source (10 to 90 percent) and t_{pd}' is the actual loaded propagation delay of the line.

If a line does not fall into the *critical line* category, i.e., it is shorter than the critical length, the line delay used in the timing budget is the one-way propagation delay of the line (t_{pd}' times the line length). If a line exceeds the *critical line length,* some allowance for transmission-line effects must be incorporated into the timing analysis unless the line is terminated in such a way that the line is stable at all loads after a one-way delay. For the general category of control and data signals, it is impractical to terminate all lines. Thus, additional time beyond the one-way line delay must be allowed for most signals to settle to acceptable logic levels. For most lines in excess of the *critical line length,* allowing five line delays will provide adequate time for transmission-line reflections to subside (see Section 11-5).

12-2-1 Transmission lines with distributed loads

Lines with distributed loads slow the propagation of signals by a factor equal to

$$t_{pd}(\text{slow down factor}) = \sqrt{1 + \frac{C_{LOAD}}{C_{LINE}}} \tag{12-1}$$

Thus, where loads are distributed, the actual propagation time t_{pd}' is given by Equation (11-9).

$$t_{pd}' = t_{pd}\sqrt{1 + \frac{C_{LOAD}}{C_{LINE}}}$$

When lines have distributed loads, t_{pd}' should be used in Equation (11-5) to determine the *critical line length.*

12-2-2 Lumped loads

When total line length is less than a *critical line length,* line loading appears as lumped.[6] In lumped load cases, transmission-line effects are not significant, but signal transition times may be slowed because of lumped capacitance loads. If the manufacturer's specified load is exceeded (50 pF for the advanced logic families and 15 pF for LS and S), the propagation times of the driving devices must be adjusted as described in Section 12-1.

12-3   Backpanel Interconnection Delays

The same device and line delay considerations must be addressed for backpanel or motherboard signal connections as for circuit board interconnections (see Section 12-2). However, most backpanel connections must be treated as transmission lines. Most backpanel interconnections will exceed the *critical line length*.[6] In addition to normal backpanel signal path interconnection distance, backpanel signal interconnection length and delay calculations should include an allowance for extender cards (which allow circuit boards to be extended out of a chassis so that test equipment can be connected for troubleshooting). All systems will have problems, so it is essential that board-to-board signals have adequate timing margins so that boards can be extended and the system tested. Most experienced designers have encountered a digital system in which the timing was so critical that the system would not function when the boards were mounted on extender boards. Such systems are difficult and expensive to troubleshoot and to evaluate during the developmental stage. When advanced Schottky or advanced CMOS devices are applied, the typical *critical line length* is 0.5 ft. Thus, the extra length of an extender board is certain to increase backpanel interconnections beyond the *critical line length*. Thus, additional time beyond the one-way line ae-lay must be allowed for unterminated backpanel signals to settle to adequate logic levels. For most backpanel lines, an allowance of five line delays will provide adequate time for transmission-line reflections to subside (see Section 11-5).

12-4   Unit-to-Unit Interconnection Timing

Timing analysis of signal interconnections between remote units (i.e., units that do not have a common backpanel or motherboard) must allow for transmission-line effects. In most cases, unit-to-unit signal paths will exceed the criteria for *critical line length* regardless of the logic family being applied. Thus, signals traversing unit boundaries must be properly terminated, or time must be allowed for signals to settle to valid logic levels (see Section 11-5).

Most unit-to-unit interconnections consist of some form of cabling using twisted-pair or shielded twisted-pair lines, coaxial cable, or flat cable. Cable manufacturers' specifications should be consulted for cable propagation times.[7]

Synchronous transfer of data between remote units is generally not practical because of the difficulty of maintaining clock alignment. If synchronous transfers are planned, some allowance for clock alignment uncertainty must be included in the timing analysis.

12-5   Examples of Worst-Case Timing Analysis

The signal path shown in Figure 12-2 is representative of signal paths commonly encountered in large synchronous systems where data and control signals flow between boards mounted on a common motherboard or backpanel. A signal originates at the output of a clocked device (flip-flop 1), proceeds through a buffer and leaves circuit board 1 to go to the motherboard. It then goes on to circuit board 2, where it goes through a buffer and then into a control input on a clocked device (flip-flop 2). A buffered register-to-register transfer, as shown, represents the simplest possible board-to-board signal-transfer circuitry. Perhaps the output buffer and receiving buffer could be eliminated, but they are required in most cases. Thus, the signal path propagation time of the example circuit is an indication of the maximum clock rate that can be achieved in a large digital system. Actual systems tend to have much more complex data paths.

In the example circuit shown in Figure 12-2, the worst-case signal propagation time is determined by the time required for a signal to

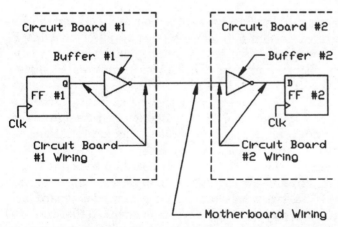

Figure 12-2 Typical board-to-board signal path.

travel from flip-flop 1 on board 1 to flip-flop 2 on board 2. The incremental signal path delays are as follows:

1. The maximum propagation delay from the clock to the output of flip-flop 1

2. The time for the signal to propagate from the output of flip-flop 1 to the input of the output buffer and the time required for the signal to stabilize at the desired logic level

3. The maximum propagation time for the output buffer

4. The time required for the signal to traverse the backpanel (motherboard) and to stabilize at the desired logic level at the input of the receiving buffer on board 2 (transmission-line effects, such as ringing, must be considered)

5. The maximum propagation time of the receiving gate (buffer)

6. The time for the signal to propagate from the output of the receiving gate to the control input of the receiving flip-flop (flip-flop 2) and stabilize at the transmitted logic level

7. The setup time required by flip-flop 2

Tables 12-1 to 12-6 show the example circuitry signal path delay for several of the more common logic families. The tables are useful for comparing logic family speed and as a guide for maximum system operating speed when a given logic family is used. The device propagation delay and setup times listed in Tables 12-1 to 12-6 are based on circuit operation over the worst-case military temperature range (-55 to $+125°C$) and power-supply variation limits (± 10 percent from nominal 5 V).

The initial device speed entries in Tables 12-1 and 12-2 for the LS and S families are worst-case data sheet values at $+25°C$ and 15-pF load. The $+25°C$ worst-case times must be multiplied by 1.5 for worst-case military operating conditions. The device timing data entries listed in Tables 12-3 to 12-6 for the ALS, FAST, AC, and HC families are data sheet -55 to $+125°C$, 50-pF load, worst-case limits. Since the ALS, FAST, AC, and HC families are specified for operation over the full military operating range, the data sheet timing specifications do not have to be further derated.

The interconnection delay calculations are based on pc board 1 and 2 track lengths of 6 in and a backpanel plus extender card interconnection length of 1.5 ft. Two nanoseconds per foot is used for board and backpanel signal track propagation delays. It is assumed that the two signal interconnections on the boards do not ring and are stable after

TABLE 12-1   Example Circuit Worst-Case Delay with LS Components[8]

1. Flip-flop 1 propagation delay	
Max value of t_{PHL} or t_{PLH} at T_A = 25°C	40.0 ns
Derating of 50% for full military temperature range operation	20.0 ns
2. Circuit board 1 track delay	
2 ns/ft × 0.5 ft +5 ns derating for additional load capacitance	6.0 ns
3. Output buffer propagation delay	
Max value of t_{PHL} or t_{PLH} at T_A = 25°C	15.0 ns
Derating of 50% for full military temperature range operation	7.5 ns
4. Backpanel propagation delay	
Assume a 1.5-ft interconnection	
5 × 2 ns/ft × 1.5 ft =	15.0 ns
5. Input gate propagation delay	
Max value of t_{PHL} or t_{PLH} at T_A = 25°C	15.0 ns
Derating of 50% for full military temperature range operation	7.5 ns
6. Circuit board 2 track delay	
2 ns/ft × 0.5 ft +5 ns derating for additional load capacitance	6.0 ns
7. Flip-flop 2 setup time	
At T_A = 25°C	20.0 ns
Derating of 50% for full military temperature range operation	10.0 ns
Total propagation delay	162. 0ns

a one-way line propagation delay. It is assumed that the interconnection between boards does ring and that five line delays are required for the backpanel interconnection to settle. All on-board signal capacitance loads are assumed to be 50 pF. Thus, the LS and S examples require compensation since those families are specified with a 15-pF load. An additional 35 pF of load can be expected to increase, *on the average,* the output delay of S devices 2 ns and of LS devices 5 ns (see Section 12-1-2, Figure 12-1). The ALS, FAST, AC, and HC families require no additional derating since they are specified with 50-pF loads.

Summary of times listed in Tables 12-1 to 12-6.   Since the circuitry and interconnections shown in Figure 12-2 represent the minimal circuitry needed for board-to-board data transfers, the resulting total signal path propagation times shown in Tables 12-1 to 12-6 are indicative of the maximum clock frequency that could be safely used for systems with similar signal paths. Table 12-7 lists maximum system

TABLE 12-2   Example Circuit Worst-Case Delay with S Components[8]

1. Flip-flop 1 propagation delay	
Max value of t_{PHL} or t_{PLH} at $T_A = 25°C$	9.0 ns
Derating of 50% for full military temperature range operation	4.5 ns
2. Circuit board 1 track delay	
2 ns/ft × 0.5 ft +2 ns derating for additional load capacitance	3.0 ns
3. Output buffer propagation delay	
Max value of t_{PHL} or t_{PLH} at $T_A = 25°C$	5.0 ns
Derating of 50% for full military temperature range operation	2.5 ns
4. Backpanel propagation delay	
Assume a 1.5-ft interconnection	
5 × 2 ns/ft × 1.5 ft =	15.0 ns
5. Input gate propagation delay	
Max value of t_{PHL} or t_{PLH} at $T_A = 25°C$	5.0 ns
Derating of 50% for full military temperature range operation	2.5 ns
6. Circuit board 2 track delay	
2 ns/ft × 0.5 ft +2 ns derating for additional load capacitance	3.0 ns
7. Flip-flop 2 setup time	
At $T_A = 25°C$	3.0 ns
Derating of 50% for full military temperature range operation	1.5 ns
Total propagation delay	54.0 ns

clock frequency for the various logic families based on the total board-to-board propagation times listed in Tables 12-1 to 12-6.

The maximum clock frequencies listed in Table 12-7 represent the best-case system clock speed for a given logic family. Note that the operating speed limit for the HC and LS families is near 5 MHz and that only the AC and FAST families allow 20-MHz operation. To build systems that will operate reliably with higher clock frequencies than those listed in Table 12-7, signal path lengths must be reduced or all board-to-board signals must be terminated to reduce settling times. Some might argue that the assumptions made are too pessimistic and that all devices in a delay chain will not exhibit worst-case propagation delays simultaneously, and that is normally true. However, if worst-case timing numbers are used for each component, a system will have some margin to compensate for those items that are overlooked, and there are always some of those.

TABLE 12-3   Example Circuit Worst-Case Delay with ALS Components[9]

1. Flip-flop 1 propagation delay	
Max value of t_{PHL} or t_{PLH} over $T_A = -55$ to $+125°C$	21.5 ns
2. Circuit board 1 track delay	
Output of flip-flop 1 to input of output buffer delay	1.0 ns
3. Output buffer propagation delay	
Max value of t_{PHL} or t_{PLH} over $T_A = -55$ to $+125°C$	14.0 ns
4. Backpanel propagation delay	
Assume a 1.5-ft interconnection	
5×2 ns/ft $\times 1.5$ ft =	15.0 ns
5. Input gate propagation delay	
Max value of t_{PHL} or t_{PLH} over $T_A = -55$ to $+125°C$	16.0 ns
6. Circuit board 2 track delay	
Output of receiving gate to input of flip-flop 2	1.0 ns
7. Flip-flop 2 setup time	
At $T_A = -55$ to $+125°C$	16.0 ns
Total propagation delay	84.5 ns

TABLE 12-4   Example Circuit Worst-Case Delay with FAST Components[3]

1. Flip-flop 1 propagation delay	
Max value of t_{PHL} or t_{PLH} over $T_A = -55$ to $+125°C$	10.5 ns
2. Circuit board 1 track delay	
Output of flip-flop 1 to input of output buffer delay	1.0 ns
3. Output buffer propagation delay	
Max value of t_{PHL} or t_{PLH} over $T_A = -55$ to $+125°C$	7.0 ns
4. Backpanel propagation delay	
Assume a 1.5-ft interconnection	
5×2 ns/ft $\times 1.5$ ft =	15.0 ns
5. Input gate propagation delay	
Max value of t_{PHL} or t_{PLH} over $T_A = -55$ to $+125°C$	7.0 ns
6. Circuit board 2 track delay	
Output of receiving gate to input of flip-flop 2	1.0 ns
7. Flip-flop 2 setup time	
Largest value of minimum t_{SH} or t_{SL} over $T_A = -55$ to $+125°C$	4.0 ns
Total propagation delay	45.5 ns

TABLE 12-5   Example Circuit Worst-Case Delay with AC Components[2]

1. Flip-flop 1 propagation delay	
Max value of t_{PHL} or t_{PLH} over $T_A = -55$ to $+125°C$	11.5 ns
2. Circuit board 1 track delay	
Output of flip-flop 1 to input of output buffer delay	1.0 ns
3. Output buffer propagation delay	
Max value of t_{PHL} or t_{PLH} over $T_A = -55$ to $+125°C$	8.0 ns
4. Backpanel propagation delay	
Assume a 1.5-ft interconnection	
5×2 ns/ft $\times 1.5$ ft =	15.0 ns
5. Input gate propagation delay	
Max value of t_{PHL} or t_{PLH} over $T_A = -55$ to $+125°C$	8.0 ns
6. Circuit board 2 track delay	
Output of receiving gate to input of flip-flop 2	1.0 ns
7. Flip-flop 2 setup time	
Largest value of minimum t_{SH} or t_{SL} over $T_A = -55$ to $+125°C$	3.5 ns
Total propagation delay	47.0 ns

TABLE 12-6   Example Circuit Worst-Case Delay with HC Components[10]

1. Flip-flop 1 propagation delay	
Max value of t_{PHL} or t_{PLH} over $T_A = -55$ to $+125°C$	55.0 ns
2. Circuit board 1 track delay	
Output of flip-flop 1 to input of output buffer delay	1.0 ns
3. Output buffer propagation delay	
Max value of t_{PHL} or t_{PLH} over $T_A = -55$ to $+125°C$	46.0 ns
4. Backpanel propagation delay	
Assume a 1.5-ft interconnection	
5×2 ns/ft $\times 1.5$ ft =	15.0 ns
5. Input gate propagation delay	
Max value of t_{PHL} or t_{PLH} over $T_A = -55$ to $+125°C$	46.0 ns
6. Circuit board 2 track delay	
Output of receiving gate to input of flip-flop 2	1.0 ns
7. Flip-flop 2 setup time	
Largest value of minimum t_{SH} or t_{SL} over $T_A = -55$ to $+125°C$	24.0 ns
Total propagation delay	188.0 ns

TABLE 12-7   Maximum System Clock Frequency for Various Logic Families

Logic family	Maximum signal path delay, ns	Maximum system clock frequency, MHz
HC	188	5.3
LS	162	6.2
ALS	84.5	11.8
S	54	18.5
AC	47	21.3
FAST	45.5	22

12-6 Summary

A great deal of care and attention to detail must be given to system timing when advanced Schottky and advanced CMOS devices are used. The optimistic typical timing specifications prominently displayed on the first page of most data sheets are of little value and should be disregarded. The parameter F_{MAX}, which is listed as the maximum toggle rate or maximum clock rate on data sheets for clocked devices such as counters, flip-flops, and shift registers, should never be used as an indication of the useful speed of a device.[1] In all high-speed applications, actual in-circuit worst-case device timing parameters must be used to establish system timing limits. System timing must allow for signal interconnection delays since interconnection delays are generally a major portion of most signal delays in high-speed systems.

Carefully follow synchronous design practices so that system timing requirements can be understood and analyzed. Structure the design so that a measure of actual operating speed can be determined. Provide a method for inserting a variable-frequency test clock so that actual system performance can be determined during development and so that individual unit performance can be verified during production. Under ideal conditions (i.e., +25°C and +5 V), a well-designed system with the margin to allow for worst-case military operating conditions and degradation with time should operate at near twice the design speed. Systems with less than 20 percent speed margin under ideal conditions, even if intended for benign conditions, present a high level of risk because of the variability of parts, actual operating conditions, and degradation of parts with age.

12-7 Signal Timing Checklist

1. Use manufacturer's worst-case military or commercial timing values when they are available for the operating conditions of interest (i.e., system temperature, load, and supply voltage levels).

2. Use the following multipliers to estimate worst-case timing values if worst-case temperature and supply variation timing values are not available:

- 1.5 to change +25°C typical to 0 to +70°C worst case
- 1.25 to change +25°C maximum to 0 to 70°C worst case
- 2.0 to change +25°C typical to −55 to +125°C worst case
- 1.5 to change +25°C maximum to −55 to +125°C worst case

3. Adjust manufacturer's maximum timing values for load capacitance in excess of that at which the timing is specified. Adjust minimum timing values when the load is less than the specified load. Use manufacturer's timing adjustment guidelines or estimate change in delay using

- 200 ps/pF × ΔC for HC
- 150 ps/pF × ΔC for LS
- 80 ps/pF × ΔC for ALS
- 50 ps/pF × ΔC for S, AC, FAST, LS, and ALS drivers

where ΔC is the actual load minus the specified load.

4. The signal line capacitance needed for timing calculations is equal to the sum of the input and output capacitance of all devices connected to the line plus the line capacitance.

Actual device input or output capacitance should be used when available. If actual input-output capacitance is not available, use the following guidelines for standard SSI and MSI parts:

- 5 pF per input
- 7 pF per output
  15 pF per bidirectional port

For VLSI parts such as PAL, gate arrays, etc., manufacturer's data must be consulted. Some PAL have 20 pF or more of capacitance per input or output.

Estimate line capacitance using

- 1.5 pF/in for welded-wire and wire-wrap interconnections
- 3 pF/in for pc board traces

5. For interconnection propagation delays use

- 1.5 ns/ft for welded-wire and wire-wrap interconnections
- 2 ns/ft for pc board traces

For cables or twister pair lines, consult manufacturers' data.

Caution: Some PVC insulated cables have very slow propagation times (greater than 3 ns/ft).

6. Allow for the additional length of an extender board or test cable in the timing analysis when appropriate (e.g., board-to-board signals).

7. Terminate long signal lines (greater than the *critical length*) with critical timing or wave-shape requirements.

■ Allow one line delay for load-terminated lines.

■ Allow two line delays for source-terminated lines.

8. Allow five line delays for unterminated long lines to settle.

References

1. *Bipolar Microprocessor Logic and Interface Data Book,* Advanced Micro Devices Inc., Sunnyvale, Calif., 1981.
2. *FACT—Advanced CMOS Logic Databook,* National Semiconductor Corp., Santa Clara, Calif., 1989.
3. *FAST—Advanced Schottky TTL Logic Databook,* National Semiconductor Corp., South Portland, Me., 1988.
4. Abramson, S., C. Hefner, and D. Powers: *Simultaneous Switching Evaluation and Testing Design Considerations,* Texas Instruments Inc., Dallas, Tex., 1987.
5. *DATABOOK, RCA High-Speed CMOS Logic ICs,* RCA Corp., Somerville, N. J., 1986.
6. Royle, David: "Transmission Lines & Interconnections, Part One," *EDN,* June 23, 1988, pp. 131–136.
7. Crouch, Ronald A.: "Choose Cable with Care to Optimize System Design," *EDN,* November 5, 1978, pp. 113–116.
8. *TTL Logic Data Book,* Texas Instruments Inc., Dallas, Tex., 1988.
9. *ALS/AS Logic Data Book,* Texas Instruments Inc., Dallas, Tex., 1986.
10. *RCA QMOS Integrated Circuit Databook,* RCA Corp., Somerville, N. J., 1985.

Reset Signals

Each digital system should have a single central master reset signal generator. The master reset signal should be buffered and distributed to all units for system initializations at power turn ON or following low-voltage transient conditions. A single reset signal source (generator) is necessary to ensure that all sections of a system are initialized at the same time and under the same conditions. Distributed reset signal generators using RC networks should not be used because of the uncertainty of their operation. Distributed reset signal generators introduce the possibility that sections of a system may not be initialized or that sections may not be initialized at the same time.

The master reset signal should be applied for a short time when system power is turned ON, or when the supply voltage V_{cc} falls below the actual level at which the devices used in the system no longer operate. It is not practical to use reset circuits with the trip point (threshold) set at exactly the minimum-rated device supply-voltage level; some allowance for system noise and an allowance for inherent accuracy limitations of practical level-sensing circuits is needed. In most applications, the trip point should be set 0.25 to 0.5 V below the rated low power-supply operating level of the system components with the highest minimum rating. Some safety margin is needed to avoid excessive tripping due to noise and to provide an allowance for error in the level sensing circuitry. Most systems designed with proper safety margins will continue to operate with supply levels slightly below rated minimum levels. However, when a system must either operate properly or shut down and be reinitialized if a low-voltage condition occurs, the reset circuit trip point must be set above the point at which the system stops operating.

For most TTL and CMOS devices with TTL input levels, the low V_{cc} operating limit is 4.75 V for commercially rated devices and 4.5 V for military rated devices. However, most CMOS devices with TTL inputs will operate at much lower voltages, but device input-output levels may not meet TTL specifications and speed is reduced. Most advanced CMOS devices with CMOS inputs have a rated low supply operating limit of 2 V, but device speed decreases significantly at reduced supply levels. Thus, in all CMOS systems operated near the speed limits of the individual components, it is best to sense a minimum supply level well above the specified low supply limit. Most systems will consist of a mixture of TTL and CMOS; thus in most cases the system reset signal generator threshold should be set near the low supply operating limit for TTL devices.

The central reset signal should be generated using an analog voltage-level sensing circuit. Time-delay circuits using RC networks should never be used to generate reset signals because of the uncertainty of their trip point (switching threshold) and the unpredictability of their response to power-supply transients or power-supply turn ON or OFF rates. Simple RC reset circuits may operate as expected in the lab, but when exposed to actual system power turn ON or OFF rates, they may not function as expected; most system operating conditions differ greatly from lab conditions. Actual system power-supply turn ON rates tend to be much slower than those of breadboard setups or test station power turn ON rates. If the power supply turns ON very slowly, simple RC delay circuits may not issue a reset signal. Furthermore, most automatic test equipment cannot test RC-generated resets that time out as a function of the application of V_{cc} to the unit under test.

The central reset signal should be buffered and distributed to each unit, board, etc., that requires a reset signal. Where the reset signal enters units or boards, the signal should be buffered so that units or boards present only a single load to the central source.

13-1   Recommended Reset Signal Generator

Figure 13-1 shows an analog comparator network configured to generate a reset signal when the supply voltage V_{cc} is below a predefined level.[1,2] A circuit such as shown in Figure 13-1, or an equivalent circuit, should be used as the central source for a system master reset signal. In contrast to crude RC networks, analog comparator circuits can accurately sense low V_{cc} conditions.

The analog voltage level sensing network shown in Figure 13-1 operates as follows: The LM193 IC analog comparator $U1$ compares the 1.22-V reference input from the LM113 voltage reference to the output of the divider network, $R1$ and $R2$.[3] When the voltage level out of the

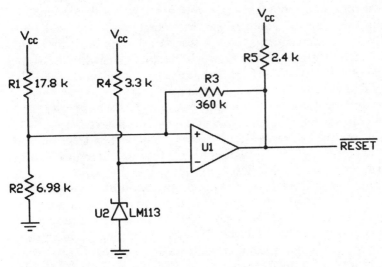

Figure 13-1   Reset signal generator using an analog comparator.

divider network is greater than the reference, the output of the comparator is a *high* TTL level. When the output of the divider network is lower than the reference, the output of the comparator is a *low* TTL level. The selection of the values of $R1$ and $R2$ controls the level of the sensed voltage, V_{cc} in this case, at which the comparator output switches. With the resistor values shown in Figure 13-1, the comparator switches states at about 4.3 V. Resistor $R3$ provides a small amount of hysteresis to prevent oscillation of the comparator when the sensed voltage is near the threshold level. Unless hysteresis is used, oscillation will occur because of the noise riding on the sensed voltage or the reference voltage. All spurious oscillations should be avoided where possible. Spurious oscillations increase system noise and may result in other system malfunctions. In noisy applications, filter capacitors should be used at the two inputs of the comparator to reduce the chance of noise triggering the comparator.

Two areas of the circuit are critical with respect to threshold accuracy: the divider network $R1$ and $R2$ and the reference $U1$. To achieve an accurate threshold level, resistors with a tolerance of 1 percent or better should be used for $R1$ and $R2$. The tolerance of the other resistors in the circuit is not critical. Reference uncertainty tends to be the largest source of error in a level-sensing circuit. Low-voltage precision integrated circuit references, such as the LM113, are required to meet reasonable threshold tolerance limits. Low-voltage zener diodes do not have sufficient accuracy for most applications. Divider networks connected to system V_{cc} or other supply voltages should not be used as a reference; normal supply-voltage levels are not precise enough and are often noisy. In all applications, the worst-case threshold limits should be determined using the worst-case tolerances of the components selected. If the results are unsatisfactory, higher-quality components should be selected. A worst-case threshold tolerance of ± 0.25 V is achievable with low-cost standard components.

It is important to select a comparator and reference that will operate predictably when the supply voltage (V_{cc}) is less than 4.5 V. In most applications, the reset circuit must be powered by the same supply that is being sensed.[4] In many digital systems, there are no other (higher) supply voltages available. In these cases, i.e., when no higher voltage supplies or independent supplies are available, it would serve no useful purpose to employ a low-voltage sensing circuit that may not work correctly when the supply level that it must sense (and be powered by) is lower than the nominal TTL power-supply operating range (4.5 to 5.5 V). The sensing circuit should be designed to operate from a low supply level of 2 to 2.5 V to a high level of at least 7 V (which is the absolute maximum upper limit for most TTL and CMOS devices—the sensing circuit should have at least as much high-voltage tolerance as the digital components). The LM193 comparator and the LM113 voltage reference used in the low-voltage sensing circuit shown in Figure 13-1 have a suitable range for such applications. The LM193 is specified for operation with a supply voltage as low as 2 V and as high as 36 V. The LM113 provides a 1.2-V reference level for a wide range of input currents. Resistor $R4$ must be selected so that the current supplied to the LM113 is adequate for the LM113 to be in regulation at a V_{cc} level much below the desired low V_{cc} threshold level.

If a central reset signal generated by a low-voltage sensing circuit must be combined with other reset signals and it is important that the system reset signal be asserted under marginal power conditions to prevent unsafe operation (e.g., false outputs, bus contention, or *write* signals to nonvolatile memories), the combination circuit must operate correctly under low-voltage conditions. Schottky TTL gates are only specified for operation above 4.5 or 4.75 V and thus are not suitable, but CMOS OR gates are available that are specified for operation with supply voltages as low as 2 V.[5] Thus, a CMOS OR gate, as shown in Figure 13-2, provides one means of combining reset signals when reset signals must be applied under low-voltage conditions.

If the only objective is to initialize a system to a known starting point and it is certain that it is not necessary to guard against unsafe conditions during power excursions, then there is no need to use special components to combine multiple-reset signals. If there are no unsafe conditions to guard against, it is of no consequence for the reset signal to be in an *indeterminate state* for a short time during and fol-

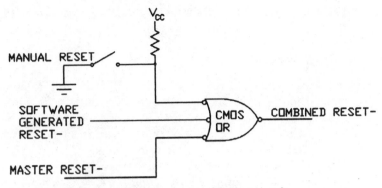

Figure 13-2   CMOS gates must be used to OR reset signals.

lowing power turn ON. In those applications the principal function of a reset signal is to initialize the system to a known starting point. When system initialization is the main function of a reset signal, it is sufficient for the reset signal to be issued a short time after power is applied. In those applications, the combination circuit does not need to function at low voltage, and as a result, no special circuitry is required to combine reset signals. For system commonality, the device or devices used should be selected from the logic family being used to implement the system.

13-2   Reset Signal Distribution

Great care must be taken in the design of the electrical and physical network that is used to distribute reset signals. Reset signals must not be degraded by excessive loading or noise coupling, and they should be buffered at a single central point in each system unit (board, chassis, rack, etc.) and redriven to each internal subunit (component, board, etc.) so as to minimize the length and load of reset signal lines. Reset signal buffers and signal buffers driving heavy loads (greater than 15 pF) should not be mixed in a common device (package). Large transient load currents may cause local power-supply droop, which will appear as negative-going spikes on the reset lines. If a *low* inactive reset signal polarity is used (in TTL systems), it is even more important that reset signal buffers do not share packages with other buffers driving heavy loads. Ground-bounce spikes on inactive *low* reset lines are more serious than on *high* inactive lines since *low* TTL signals have less noise margin than *high* signals. However, in spite of the desirability of isolating reset signals, board space limitations often make it impractical to isolate reset signal buffers. If reset signal buffers must share a package with other active buffers, the key to avoiding system operational problems is to ensure that the other buffers are very lightly loaded (the ac load is of most concern).

13-3 Reset Signal Phasing

Since most reset (set, clear, or preset) inputs on TTL or CMOS logic devices are asynchronous and thus sensitive to noise spikes, it is important that the reset signal have a *high* level of noise immunity. Take advantage of the higher, *high*-level noise margin of Schottky TTL devices or CMOS devices with TTL input levels. The *inactive* level of all TTL-level reset signals that must pass through noisy areas, such as motherboards, should be the *high* level. Most reset (clear) or preset (set) inputs on most TTL devices take advantage of the higher noise margin of the *high* state at the device level by requiring an active *low* input level for resetting (or presetting). In most applications, reset signals should be phased to be in a *high* inactive state throughout the interconnection system. To further enhance the noise margin of *high inactive* reset signals, use 1-kΩ pull-up resistors in each electrically isolated reset signal line segment to provide additional noise margin. A pull-up resistor will ensure that an inactive reset line is near 5 V, which will provide an additional 1.5- to 2-V noise margin under typical conditions and near 2.5 V under worst-case conditions (the worst-case minimum output *high* level of an advanced Schottky buffer without a pull-up resistor is 2.5 V). Pull-up resistors also serve to keep open reset lines (a line not connected to a source) in the *inactive* state. For example, pull-up resistors are often necessary at major interfaces to ensure that reset signals are not active during checkout or test when a unit may not have all signal sources present.

13-4 Reset Signal Loading

It is important to limit the number of loads on reset lines. There is often a temptation to connect an excessive number of loads to reset lines since reset signals are "slow" dc signals. The rationale is that AS or FAST devices are rated to drive 30 or more dc loads and speed is of no concern. The number of loads should be limited to about 10 for practical purposes. Troubleshooting a shorted signal line that runs to a great number of loads can be very time-consuming and difficult. Also, driving a large number of loads results in very long signal runs which increase the possibility of coupled noise. Since reset inputs are asynchronous, limiting noise on reset lines is of prime importance.

13-5 Reset Signal Timing

Reset signals should be asynchronously applied and synchronously removed.

13-5-1 Asynchronously apply reset signals

Reset signals should be asserted through direct circuit paths that have no clocked elements in the paths. The reset signal should always be asynchronously applied. The presence of the system clock should not be a condition for resetting the system. Under fault conditions, such as the absence of the system clock, all systems must have the capability of being reset to a benign state where there is no bus contention or false outputs that could initiate unsafe conditions.

13-5-2 Synchronously remove reset signals

In contrast to being asynchronously applied, reset signals should be removed synchronously to ensure that all clocked elements are properly initialized on the same clock edge. If a reset signal is removed from clocked devices, such as a multiple-stage counter, in an asynchronous manner, there is no assurance that all of the clocked elements will react uniformly.[6] Because of the metastable phenomenon (see Chapter 8), some stages may not remain reset. If some stages remain reset and some do not, the system may not react as expected. Two logic configurations that provide asynchronously applied and synchronously removed reset signals are shown in Figure 13-3.

The circuits shown in Figure 13-3 operate as follows:

Circuit a: When the RESET signal goes *low,* the LOCAL RESET signal is asynchronously activated through the AND gate *G*1. When RESET goes *high*, the transition of LOCAL RESET to a *high* level is delayed until *F*1 is clocked *high*.

Circuit b: When RESET goes *low,* LOCAL RESET is asynchronously activated through the asynchronous reset input of *F*2. When RESET goes *high,* the transition of the output of *F*2, LOCAL RESET, to a *high* level is delayed until the next clock edge.

Note that either circuit (in Figure 13-3) may glitch as a result of metastable behavior if the asynchronous RESET goes away at the same time that the active clock edge occurs, but that should not cause a problem in most applications since the LOCAL RESET signal should be stable at the other clocked elements on the board. If a glitch could cause a problem, RESET can be double-registered, thus removing the possibility of LOCAL RESET glitching.

In most applications, the reset control circuitry can be located at the central source of the reset signal or in a central location in each given unit. However, when a system is operating near the upper end of the speed capability of the logic family being used, reset signals may need

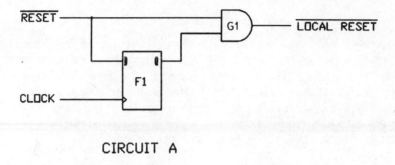

CIRCUIT A

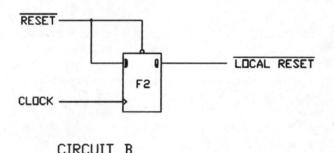

CIRCUIT B

Figure 13-3 Two circuits for asynchronously applying and synchronously removing reset signals.

to be synchronized at the local circuit board level as a result of clock phasing uncertainty. If a synchronizing circuit must operate, i.e., supply an active reset signal, during undervoltage conditions, the circuit should be implemented with advanced CMOS devices.

13-6 Summary of Reset Signal Generation and Distribution Techniques

1. Use a single central reset signal source.

2. Generate the central reset signal using an analog comparator level sensing circuit.

3. Do not use RC networks to generate reset signals.

4. Do not overload reset signal drivers or allow reset signal buffers to share packages with buffers driving large loads.

5. Buffer reset signals where they enter units, boards, etc., so as to present only a single load to the central source.

6. Pull-up reset signals where they enter units, boards, etc., so that reset will not be active when the normal source is not present.

7. The phasing of TTL-level reset signals should be such that reset lines are *high* in the inactive state.

8. Pull-up *high* inactive TTL-level reset signals for extra noise margin.

9. Reset signals should be asynchronously applied and synchronously removed.

10. Use CMOS logic components with low supply-voltage operating capability throughout the reset signal path when it is important for reset signals to be active to prevent unsafe conditions during transient power-supply conditions.

References

1. Frederiksen, Thomas M.: *Intuitive IC Op Amps,* National Semiconductor Corp., Santa Clara, Calif., 1984.
2. *Linear Applications Databook,* National Semiconductor Corp., Santa Clara, Calif., 1986.
3. *Linear Databook 1,* National Semiconductor Corp., Santa Clara, Calif., 1988.
4. Austin, W. H., J. E. Buchanan, and C. W. Nelson: "Precise Voltage Level Detector," *Digital Design,* August 1973, p. 30.
5. *FACT—Advanced CMOS Logic Databook,* National Semiconductor Corp., Santa Clara, Calif., 1988.
6. Chaney, Thomas J.: "Measured Flip-Flop Responses to Marginal Triggering," *IEEE Trans. Computers,* Vol. C-32, No. 12, December 1983.

Unused Inputs

Unused inputs on powered devices must never be left open or floating; a valid *high* or *low* logic level is required at all times.[1] Valid input logic levels are required to ensure that input circuits do not oscillate and upset device operation or increase system noise levels. Unused inputs on CMOS devices are of particular concern; CMOS devices overheat and may be destroyed if inputs are left open.

14-1 Unused TTL Inputs

Unused TTL inputs fall into two basic categories:

1. Unused TTL inputs requiring a logic *low* should be tied directly to ground. The use of a pull-down resistor may shift the input level above a logic *low* level. If a pull-down resistor must be used for some reason, such as testability, it must be sized so as to ensure that minimum *low* input level requirements are not exceeded. *Don't care* inputs are generally tied to a logic *low* level for simplicity of wiring.

2. Unused TTL inputs that require a logic *high* fall into two classes.

a. Those that can be tied directly to V_{cc}.

Unused inputs on AS, LS, ALS, and FAST devices, or other TTL devices that have real diode inputs, can be tied directly to the local $+V_{cc}$ supply voltage. No electric-circuit connection rules are violated in doing so. However, current-limited static *high* input levels are recommended rather than direct connections to V_{cc}. Current-limited *highs* reduce the chance for damage due to accidental shorts and enhance testability. Pull-up (current-limiting) resistors or inverters that have their inputs tied to ground can serve as sources of current-limited static *high* levels (normal loading limits apply). The use of pull-up resistors is the most desirable arrangement.

b. Those that require current-limiting.

Standard TTL and S logic families that have multiple-emitter inputs require the use of a current-limited source of a static *high*-input level. Pull-up resistors connected between the supply and the unused inputs needing to be pulled up can serve as a current-limited logic *high*, or a *high* output from an inverter with its input grounded can be used.

CAUTION: **Never connect standard TTL or S inputs directly to** V_{cc}. **No exceptions**

When pull-up resistors are used as *high* input sources, the number of inputs connected to a given pull-up string must not load the pull-up resistor to the extent that the voltage drops below minimum input *high* requirements. As a practical matter the number of loads should be limited to well below the number that could be driven based on the total input current load. The number of loads should be limited to simplify troubleshooting (for example, of a shorted line) and to minimize line length so as to minimize the possibility of coupled noise. As a rule of thumb, up to 10 inputs can be safely pulled up by a 1-kΩ resistor.

14-1-1 Current-limited pull-up sources

Current-limited static *high*-level pull-up sources are required for TTL devices with multiple-emitter inputs.

Input pull-up resistors are required for certain TTL logic families and not for others, depending on whether the input structure consists of real diodes or multiple-emitter transistors. If the input circuitry consists of a multiple-emitter transistor, a current-limiting resistor is required. If the input network consists of real diodes, current-limiting is not needed.

The newer TTL logic families, AS, ALS, and FAST, as well as the older LS logic family, have input circuits implemented with real diodes. An example of a diode input structure (that of a 2-input LS NAND gate) is shown in Figure 14-1. Either standard silicon diodes or combinations of silicon and metal-silicon diodes are used, but in either case the diodes used are real diodes that do not interact with one another in unexpected ways.

When using devices with real diode inputs, there is no reason from an electric-circuit standpoint to use an external current-limiting pull-

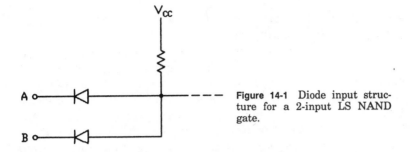

Figure 14-1 Diode input structure for a 2-input LS NAND gate.

up resistor. Devices with real diode inputs have a maximum *high* input voltage limit equal to the V_{cc} supply-voltage limit (7 V for most TTL devices). Thus, the power-supply voltage range is not restricted when diode inputs are directly connected to the power supply. However, that is not the case with multiple-emitter inputs.

Multiple-emitter inputs are only specified for an absolute maximum input level of 5.5 V.[2] If multiple-emitter inputs are connected directly to the supply, positive power-supply voltage excursions must be limited to a maximum of 5.5 V. However, it is difficult to limit power-supply excursions to 5.5 V. Thus, when standard TTL or S devices with multiple-emitter inputs are used, the safest approach is to current-limit inputs. Under no circumstance should production hardware using standard TTL or S logic devices be designed without some form of input current-limiting.

14-1-2 Multiple-emitter inputs

Figure 14-2 shows a typical TTL or S-TTL input structure consisting of a multiple-emitter transistor. Multiple-emitter devices should never be viewed as simple diodes. Multiple-emitter transistors have several current paths that may not be considered on first inspection. The various input emitters are in effect lateral transistors with finite gain. Thus, current can flow between emitters, depending upon the external circuit. Current can also flow in the inverse transistor formed

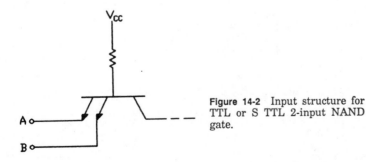

Figure 14-2 Input structure for TTL or S TTL 2-input NAND gate.

by the emitters and the normal collector. Both paths must be considered and the current limited in both.

When viewing Figure 14-2, or any similar multiple-emitter input structure, it must remembered that the assignment of the emitter and collector is not absolute. A symmetrically constructed *npn* or *pnp* transistor functions equally well regardless of which *n* terminal of a *npn* transistor or which *p* terminal of a *pnp* transistor is considered the emitter or the collector. The junctions in a multiple-emitter input transistor are not symmetrical, but the various possible combinations of junctions will function, to some degree, as transistors with gain regardless of which terminals are considered emitters or collectors. Thus, input A or input B (in Figure 14-2) will function as a collector or as an emitter depending upon the arrangement and polarity of the external connections. The magnitude of the input-current flow when the inputs of a multiple-emitter input are connected as shown in Figure 14-3 depends upon the physical spacing between the multiple emitters and the basic characteristics of the particular integrated circuit. In almost all cases, multiple-emitter inputs will have some gain. Chip designers try to design multiple-emitter input circuits so that the lateral transistor gain is very low, but chip size must also be minimized. When the alternatives are weighed, chip-size constraints usually force emitters to be located close enough to each other so that most composite input structures have some gain.

Thus, when a device has a multiple-emitter input structure, care must be taken to externally limit all input current paths. To do otherwise sacrifices device and system reliability. For example, if the A input of the device shown in Figure 14-3 is tied directly to V_{cc} and the B input is tied to ground, a transistor circuit is formed that has a source of base drive current and no current-limiting in either the collector or emitter. Transistor circuit designers will recognize the circuit configuration as a very dangerous arrangement. Transistors usually do not last very long with no current-limiting in either the collector or emitter. Yet, that is the circuit configuration when current-limiting

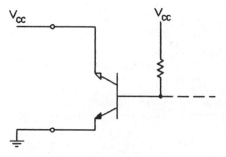

Figure 14-3 Dangerous connection.

resistors are not used to pull-up standard TTL or Schottky TTL inputs.

In addition to the transistors formed between input emitters, emitters with *high* inputs operating in conjunction with the common collector also form inverse transistors (see Figure 14-4).

It is difficult to predict under what conditions excessive input current may occur in the various inverse or parasitic lateral transistors. Destruction usually does not occur under benign conditions, but inputs tied directly to the power supply are in great danger during power-supply transients or overvoltage conditions.

In addition to increasing the possibility of device failure, an unused emitter with an unlimited input source can cause a logic-state failure in an associated used emitter. Excessive current may flow in both the unused and the used emitter when the used emitter is at a logic *low* level. The unexpected current flow in the used emitter, which can be much in excess of the normal input current, may exceed the current sink capability of the device driving that input. The possible result is that a valid logic *low* level can no longer be maintained by the driver and a logic error occurs.

In summary, always use a pull-up current-limiting resistor or some method of current-limiting to provide a *high* level for unused inputs for TTL logic elements that have multiple-emitter transistor input stages—there should be no exceptions.

14-2   Unused CMOS Inputs

Unused CMOS inputs must never be allowed to float. Open CMOS inputs may float into the undefined input region between a valid *high* and *low* logic level where both the pull-up *p*-channel and the pull-down *n*-channel complementary FETs in the input stage turn ON. With both pull-up and pull-down FETs ON, a low-impedance path is created between V_{cc} and ground. For example, Figure 14-5 shows a CMOS inverter with an open input and the resulting equivalent cir-

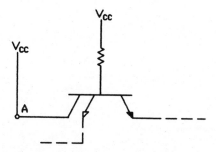

Figure 14-4 Inverted transistor formed by an input emitter and the common collector.

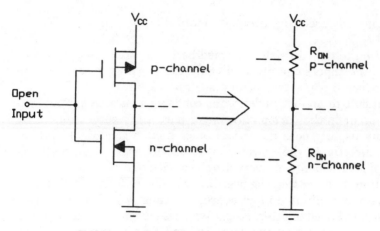

Figure 14-5   CMOS input stage and open input equivalent circuit.

cuit, which consists of two resistors between V_{cc} and ground. The resistors represent the ON impedance of the two FETs. Since the ON
impedance of the FETs used in advanced CMOS devices is relatively
low, internal paths initiated by open inputs can cause excessive internal device currents.[3] The typical profile for CMOS device supply current versus input voltage is shown in Figure 14-6.[4-6] Excessive internal current that results from open inputs can cause overheating and
destruction of parts. Even if the open input condition only lasts for a
very short time, device temperature is increased as a result of extra
current flow. High device temperature increases the susceptibility of
CMOS parts to latch-up, which leads to additional high-current flow
and a greater possibility of device destruction. The time a part can en-

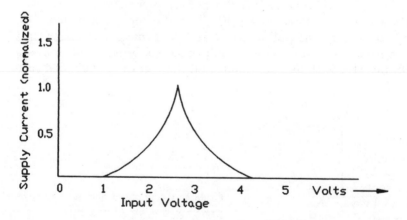

Figure 14-6   Supply current versus input voltage for CMOS buffers.

dure with an open input depends on the internal impedance of the device and other considerations such as how deep into the internal circuitry the intermediate logic level propagates. In general, floating input conditions should not be allowed to last longer than a few microseconds. However, some CMOS parts have specified minimum rise time requirements that translate into more stringent requirements. In all applications, when CMOS logic devices are used, minimum rise time requirements must be determined and adhered to in the design.

Unused CMOS inputs on used devices should be pulled *high* or *low* as required for proper logic device operation. Unused inputs on unused devices can be connected to the most convenient level. Theoretically all advanced CMOS device inputs can be connected directly to V_{cc} or ground; no current-limiting is required from an input circuit operation standpoint. However, it is best to use a pull-up or pull-down resistor to facilitate testing (a *high*- or *low*-input level can be injected) and to prevent the possibility of latch-up due to transient input current injection during power turn ON or under other transient supply conditions. Latch-up due to transient currents in inputs connected to low-impedance sources (such as V_{cc} or ground) is a known possibility with some of the older CMOS logic families. Latch-up should not be a problem when advanced CMOS devices are used, but it is safer to current-limit connections to unused inputs.

Unused inputs that are often overlooked and not pulled up or down include bidirectional ports, unused PAL or PLD inputs, and PAL outputs that serve as feedback paths. No CMOS device terminals can be allowed to float to an indeterminate logic level, even for a short length of time.

Circuit board or system interface lines that connect to CMOS inputs must have pull-up or pull-down resistors so that inputs do not float in the absence of the normal source of the signals. Often during board or system test all input sources are not present, and without pull-ups or pull-downs those CMOS devices with open input lines will be damaged.

14-3 Pull-up Resistors

For most TTL or CMOS applications where pull-ups are needed, 1-kΩ resistors are suggested.[1] Higher values increase the chance for noise pick-up (due to higher source impedance) and the chance for problems due to board leakage currents or defective devices with excessive input current. Loading of pull-up resistors (i.e., the number of inputs pulled up by each pull-up resistor) should not approach the theoretical limit which is

$$V_{cc\,min} - (I_{in\,total})(R_{pull\text{-}up}) \geq V_{in\,high\,min}$$

Since TTL *high* input currents are very low (20 μA) and CMOS even lower (1 μA), many inputs can be pulled up with one pull-up resistor without violating minimum *high* input level requirements. However, it is generally not wise, nor practical, to tie too many points together. Troubleshooting a shorted line tied to a great number of points can be very difficult, and the longer the line, the greater the chance for coupled noise. Thus, for practical considerations, the maximum number of inputs pulled up by one resistor should be limited to 10.

14-4   Additional Sources of Current-Limited Pull-up or Pull-down Voltages

Other connections that can be used to tie off unused inputs include:

1. Tie unused inputs that require a *high* input to the *high* output of an inverter that has its input tied to ground.

2. Use the output of an inverter or noninverting buffer that has its inputs tied to the appropriate level when a current-limited *low* level is required.

3. Tie unused inputs to used inputs that are functionally the same (on the same device).

14-5   Paralleling Inputs

When unused inputs are tied to used inputs (on the same device) that are functionally the same, the logical operation of the device is not affected and the unused inputs are prevented from floating. However, paralleling inputs can degrade dynamic performance and may affect dc loading in TTL applications. The capacitance load of the input signal is increased, which can adversely affect high-frequency performance, and noise margin may be reduced. Paralleling inputs parallels input capacitance and increases the capacitances between the input and the internal structure of the device. The increased capacitance increases the possibility of noise on the input signal being injected into the internal circuitry of the device and upsetting it. Thus, paralleling inputs is not recommended for high-speed devices (and specifically should be avoided for LS devices).[7]

14-6   Summary of Unused Input Connection Requirements

1. Unused inputs on used or spare devices must not be allowed to float to prevent oscillation and excessive system noise.

2. Inputs to CMOS devices must never be allowed to float to prevent excessive internal feedthrough current flow.

3. Board or system interface connections to CMOS inputs must have pull-up or pull-down resistors to prevent inputs from floating when normal input sources are not present.

4. Static pull-up voltage sources for unused TTL inputs must be current-limited if the input circuit being pulled *high* has multiple-emitter inputs.

References

1. Sokal, Nathan O.: "Designer's Guide to PC- Board Logic Design—Part 1," *EDN,* November 13, 1986, pp. 253–262.
2. *TTL Logic Data Book,* Texas Instruments Inc., Dallas, Tex., 1988.
3. Cox, Gerald C.: "Impedance Matching Tweaks Advanced CMOS IC Testing," *Electronic Design,* April 1987, pp. 71–74.
4. *Advanced CMOS Logic Designer's Handbook,* Texas Instruments Inc., Dallas, Tex., 1987.
5. *FACT—Advanced CMOS Logic Databook,* National Semiconductor Corp., Santa Clara, Calif., 1989.
6. *DATABOOK, RCA High-Speed CMOS Logic ICs,* RCA Corp., Somerville, N. J., 1986.
7. *Bipolar Microprocessor Logic and Interface Data Book,* Advanced Micro Devices Inc., Sunnyvale, Calif., 1981.

Appendix

A-1 Conversion Factors

1 foot (ft) = 30.48 cm
1 foot (ft) = 0.3048 m
1 inch (in) = 2.54 cm
1 mil = 2.54×10^{-3} cm
1 mil = 10^{-3} inches (in)

A-2 Definition of Symbols and Acronyms

AC   advanced CMOS logic

ACT   TTL-compatible advanced CMOS logic

ALS   advanced low-power Schottky TTL logic

ALU   arithmetic logic unit

AND   a logic circuit whose output is a 1-state only when every input is in the 1-state

AS   advanced Schottky TTL logic

AWG   American wire gage

b   distance between reference planes in printed circuit boards

b/s   bits per second

C   capacitance or designates a capacitor

ΔC difference in specified and actual load capacitance

C_d decoupling capacitance

C_{IN} input capacitance

C_L or C_{LOAD} load capacitance

C_{LINE} line capacitance

CLK   designates a clock signal

CLK–   designates an inverted clock signal

C_m mutual capacitance between lines

CMOS   complementary metal oxide semiconductor

C_{pd} internal device capacitance used for power calculations *d*   diameter of wire conductors

D   designates a diode

D   designates the *drain* terminal of a field-effect transistor

DIP   dual-in-line package

ECL   emitter-coupled logic

EIA   Electronic Industries Association

f   frequency (Hertz)

F   toggle frequency

F_{MAX} maximum toggle rate of clocked logic devices

F   FAST advanced Schottky TTL logic

FAST   Fairchild advanced Schottky TTL logic

FACT   Fairchild advanced CMOS logic

FCT   fast CMOS TTL-compatible logic

FET   field-effect transistor

G   designates the gate terminal of a field-effect transistor

h   height of a conductor above a reference plane

H   high-speed TTL logic

HC   high-speed CMOS logic

HCT   TTL-compatible high-speed CMOS logic

I_b base current

I_c collector current

I_{cc} device supply current

I_D drain current

I_f diode forward current

I_{IH} rated input *high* current

I_{IL} or $I_{in\ low}$ rated input *low* current

$I_{in\ total}$ total input current

I_L load current

I_n pull-down current

I_o output current

I_{OH} rated output *high* current

I_{OL} rated output *low* current

I_p peak current

I_{SC} short-circuit output current

ΔI change in current

IC   integrated circuit

K_c capacitive coupling coefficient

K_L inductive coupling coefficient

l unit length of a transmission line

l length

L   low-power TTL logic

L designates an inductor

L_p package pin inductance

L_m mutual inductance between lines

L_s source inductance

LCC   leadless chip carrier

LS   low-power Schottky TTL logic

LSI   large-scale integrated circuit

MOSFET   metal-oxide semiconductor field-effect transistor

MSI   medium-scale integrated circuit

n semiconductor material with an excess of electrons

N number of turns in an inductor

NAND   a logic device whose inputs must all be in a 1-state to produce a 0-state output

NOR   a logic device where any one input or more having a 1-state will yield a 0-state output

OR   a logic device where any one input or more having a 1-state is sufficient to produce a 1-state output

p semiconductor material with a deficiency of electrons

P_d dynamic power

PAL   programmable array logic

pc   printed circuit

PLA   programmable logic array

PLCC   plastic leaded chip carrier

PROM   programmable read-only memory

Q designates a transistor

Q_i initial charge

Q_f final charge

r radius

r_1 inner radius of a circular section of a power or ground plane

r_2 outer radius of a circular section of a power or ground plane

R resistance or designates a resistor

R_L load resistance

R_o or R_{out} output resistance

R_{ON} bipolar transistor collector-to-emitter or field-effect transistor drain-to-gate ON resistance

$R_{pull\text{-}up}$ pull-up resistor

R_S source resistance

RAM   random-access memory

ROM   read-only memory

S   Schottky TTL logic

S   designates the *source* terminal of a field-effect transistor

SCR   silicon-controlled rectifier

SSI   small-scale integrated circuit

t time (seconds)

t, w thickness and width of printed circuit board conductors

t_f fall time

t_H or t_{HOLD} the time that a signal to a clocked device must be stable after application of the active clock edge

t_{HL} or t_{PHL} the propagation time from an input change to an output *high*-to-*low* transition

t_l coupled length of two signal lines

t_{LINE} propagation time of the line

t_{LH} or t_{PLH} the propagation time from an input change to an output *low*-to-*high* transition

t_p propagation delay of a designated length of conductor

t_{pd} intrinsic propagation delay of the media

t_{pd}' effective propagation delay of loaded conductor

t_r rise time

t_S or t_{SETUP} the time that a signal to a clocked device must be stable before the arrival of the active clock edge

T_A ambient temperature

TTL   transistor-transistor logic

U   designates an integrated circuit

UL unit load

V_B voltage amplitude of backward crosstalk

V_{BE} bipolar transistor base-to-emitter voltage

V_{cc} positive device supply voltage

$V_{cc\ min}$ minimum rated supply voltage

V_{CE} bipolar transistor collector-to-emitter voltage

$V_{CE\ (sat)}$ bipolar transistor collector-to-emitter voltage when fully ON

V_f final voltage or diode forward voltage

V_F voltage amplitude of forward crosstalk

V_{IL} rated input *low* voltage

V_{IH} rated input *high* voltage

V_L voltage at the load

V_{loss} voltage drop across a ground or power plane

V_o or V_{out} output voltage

V_{OL} rated output *low* voltage

V_{OH} rated output *high* voltage

V_{os} ground offset voltage

V_s source voltage

V_x voltage at a distance x on a transmission line

ΔV change in output level

ΔV signal swing

ΔV_S driving signal transition amplitude

w width

z height

Z_o characteristic impedance

Z_o' effective characteristic impedance

Greek symbols

ϵ relative dielectric constant

ϵ_o dielectric constant (permittivity) of free space (8.85×10^{-12} F/m)

ϵ_r effective dielectric constant

ρ resistivity (the resistivity of copper is $1.724 \times 10^{-6}\ \Omega \cdot$ cm at 20°C)

ρ_L load reflection coefficient

ρ_s sheet resistance

ρ_S source reflection coefficient

ϕ magnetic flux (webers)

A-3 Trademarks

FAST is a registered trademark of National Semiconductor Corporation.

FACT is a registered trademark of National Semiconductor Corporation.

PAL is a registered trademark of Advanced Micro Devices, Inc./ Monolithic Memories, Inc.

Glossary

Advanced complementary metal oxide semiconductor (CMOS) logic devices or families Advanced-performance logic devices or family of devices fabricated with state-of-the-art CMOS processes.

Advanced Schottky transistor-transistor logic (TTL) devices or families Advanced-performance bipolar logic devices or family of devices fabricated with state-of-the-art bipolar processes.

AND gate A logic circuit whose output is a 1-state only when every input is in the 1-state.

Asynchronous operation The completion of one operation triggers the next. Operations *do not* occur in step with a clock.

Asynchronous inputs or asynchronous interfaces Signals with no fixed frequency or phase relationship to the receiving device or system clock signal.

Asynchronous or dc SET or RESET inputs SET or RESET inputs on storage elements which are level sensitive, i.e., activated by a level rather than a level transition.

Backpanel A panel used for mechanical support of several interconnected circuit boards and support of the board-to-board interconnections.

Backplane A power or ground plane that is an integral portion of a backpanel.

Balanced differential communication Transmission of digital data on a pair of lines (wires) using a true and a complementary signal pair, i.e., each signal of the pair is always the inverse of the other. In a balanced differential pair, currents tend to be equal but opposite in direction so that the net current flow at any point is small.

Bandwidth The frequency at which the gain of the device or network is 3 dB below its low-frequency value.

Bilevel signals Signals with two defined levels.

Bipolar device A semiconductor device whose operation depends on the flow of both holes and electrons.

Bipolar transistor A three-layer semiconductor device whose operation depends on the flow of both holes and electrons.

Bistable   A logic device with two stable states.

Buffer   A device used to provide extra drive and to isolate low drive devices from heavy loads.

Bulk-decoupling capacitor   A capacitor located near the power entrance points of a circuit board or larger unit used to supply low-frequency transient supply-current needs and to help suppress transmission of internal noise.

Bypass capacitor   See decoupling capacitor.

Characteristic impedance (Z_o)   The apparent real impedance (resistance) of a transmission line when a signal with a sufficiently high frequency content is applied.

Circuit model or equivalent circuit   A functional representation of a circuit for a limited set of conditions using simpler and more easily understood circuit elements.

Clock   The source or the periodic signal used to synchronize synchronous systems. The periodic signal applied to clocked elements such as flip-flops, counters, etc., to activate logical operations.

Clock frequency   The clock repetition rate.

Clock period   The time interval between two active clock edges. The inverse of the clock frequency.

Clock signal   The periodic signal applied to clocked elements such as flip-flops, counters, etc., to activate logical operations.

Clock switching edge   The clock edge transition that initiates state changes of clocked elements.

Clocked device or element   A digital circuit that requires a clock input to change states.

Commercial devices or parts   Digital devices rated to operate over the 0 to +70°C temperature range and ±5 percent power-supply variations. Part numbers for commercial rated devices in many of the logic families start with 74 XXX.

Common-mode voltage   The absolute voltage present at differential inputs (of, for example, a line receiver) with respect to the device ground reference.

Complementary signal   A signal that is *high* when a related signal is *low* or is *low* when a related signal is *high*.

Critical line length   The line length that is equal to one-half the rise time of the signal divided by the actual loaded propagation delay of the line.

Crosstalk or cross coupling   Undesirable signal coupling between adjacent or nearby signal lines.

Daughterboard   A circuit board that plugs into a motherboard (connector).

Derating or drive derating   A safety factor applied to the current drive rating specified by logic device manufacturers to ensure that a device will function in a system over a long lifetime and under the extremes of the expected operating conditions.

Decoupling capacitor   A capacitor used to supply the transient switching currents associated with digital circuits.

Destination or load termination   A load impedance that has a prescribed relationship to the interconnecting line impedance (Z_o).

Device or digital device   A digital circuit, generally an integrated circuit.

Differential line driver   A line driver with a true and a complementary output used for balanced digital data transmission.

Differential line receiver   A line receiver used in balanced digital data transmission applications that requires a true and a complementary input signal.

Differential signals   A pair of signals, one of which is always the complement of the other.

Digital circuit   A circuit that switches between two levels and makes logical decisions, performs logic operations, or stores data.

Digital data   Data conveyed by means of bilevel signals.

Driver   A device used to provide extra drive and to isolate low drive devices from heavy loads.

Dynamic loading   The capacitance and inductance associated with a digital signal line, i.e., the capacitance and inductance that must be driven when a digital signal switches states.

Dynamic power dissipation   The power dissipated charging and discharging the internal and load capacitance associated with a given device or functional unit.

Edge triggered   Activation of a digital circuit by the edge of a pulse rather than the level of the signal.

Element or logic element   A digital circuit, generally an integrated circuit.

Equivalent circuit   A functional representation of a circuit for a limited set of conditions using simpler and more easily understood circuit elements.

Fan-out   The number of logic device inputs a given output is capable of driving while maintaining correct logic levels.

Feedthrough current   The transient current that flows through a device, i.e., from the supply voltage to ground when a device switches states.

Ferrite bead or shield bead   A bead made of ferrite material used to absorb high-frequency energy in fast signal transitions. Ferrite beads located at the signal source help minimize transmission-line ringing and overshoot.

Flip-flop   A circuit having two stable states and the ability to change from one to the other on the application of external signals.

Gate   A combinational logic circuit.

Ground bounce   Transient reference (ground) shifts caused by transient current flow to ground being impeded by the inductance of the device ground pin or other connections between the device and the system reference level.

Hold time (t_h) The time interval that a signal on an input pin of a clocked logic device must be retained after an active (triggering) clock transition.

Inactive state or level   The logic state or level that does not initiate action or cause an operation to occur.

Indeterminate level   A signal level that does not meet the defined requirements for a logic *high* or *low* level.

Indeterminate state   Where the conditions of the input signals, reference, or power to a device are such that the response of the device is undefined. For example, if the power-supply voltage to a device is less than the specified minimum operating level, the device may not function.

Interconnection   The wire, printed circuit board trace, or other physical means used to connect two or more electric circuits.

Input level   The actual voltage at an input terminal of a device.

Input-output circuits   The circuitry on the boundary (interface) of a device, circuit board, or unit, i.e., the circuitry that connects to other devices, boards, or units.

Integrated circuit (IC)   An electronic device in which all of the components are fabricated on a single piece of semiconductor material.

Interface   The electrical boundary of a device, system, or subsystem.

Interface circuitry   The circuitry used at device, system, or subsystem boundaries.

Inverter   A logic device that changes the input logic level from a *high* to *low* or *low* to *high*, i.e., a circuit whose output is the opposite of the input.

Latch   A flip-flop that is level-controlled rather than edged-triggered.

Latch-up   A disruptive and possibly destructive low-impedance condition that occurs in CMOS devices when parasitic SCRs are triggered ON by extraneous substrate currents. Once triggered ON, the parasitic SCR remains ON until power is removed from the part.

Level translator   A circuit for interfacing forms of logic having different logic levels.

Line capacitance   The capacitance associated with a given signal interconnection, i.e., wire, pc board trace, or other physical means of interconnection.

Line driver   A digital circuit suitable for driving long lines.

Line receiver   A digital device used to receive signals from long lines.

Load capacitance   The total capacitance of a signal interconnection, or in some transmission-line equations, the capacitance of discrete device inputs

and outputs connected to a signal line, but not including that of the interconnecting signal line.

Load or destination termination A load impedance that has a prescribed relationship to the interconnecting line impedance (Z_o).

Local decoupling capacitor A capacitor located near the power pins of a device that supplies the transient supply current needed to charge stray and
load capacitance when the device switches states.

Logic family A group of logic devices that are fabricated with a common
semiconductor technology and that have similar electrical characteristics, i.e.,
speed, power, etc.

Logic *high* level The most positive of the two logic levels.

Logic *low* level The most negative of the two logic levels.

Lumped capacitance load Capacitance within a distance equivalent to less
than one-half the rise time of the signal.

Master or central reset signal A signal used to initialize an entire digital
system to a known starting condition.

Master-slave A binary element containing two independent storage stages
with a definite separation of the clock function (edges or levels) used to enter
data into the master and transfer it to the slave.

Matching impedance A network used to match the source or load impedance
to the impedance of the interconnecting line.

Memory A digital device capable of storing a number of bits of digital data
or digital words.

Metastable An unknown or unstable output condition that can occur when
the inputs to a clocked device do not meet the required timing relationships or
levels.

Military device, part, or military rated device A digital device rated to operate over the -55 to $+125°C$ temperature range and $±10$ percent power-supply
variations. Part numbers for most military rated advanced TTL or CMOS devices start with 54 XXX.

Monostable multivibrator (one-shot) A digital circuit with two output
states, one of which is stable and the other temporary, which can be triggered
into the temporary state for a period of time determined by an associated *RC*
network.

Motherboard A large board that serves as the mechanical support for daughter circuit board connectors and that provides the interconnections between
the daughter boards.

Multilayer printed circuit (pc) board A board used to interconnect electric
circuits with more than two layers of etched interconnections determined in
the manufacturing process by printed masks.

Multiple-emitter transistor A bipolar transistor with more than one emitter.

Such transistors are typically used to implement the input stage of standard (original) and Schottky TTL logic circuits.

NAND gate   A logic device whose inputs must all be in a 1-state to produce a 0-state output.

Noise immunity   A measure of how good a circuit is at rejecting extraneous signals.

Noise margin   The amount of extraneous voltage that a signal can tolerate before the signal is no longer recognized as the intended logic level.

Nonvolatile memory   A memory device , i.e., a digital data storage device, not requiring continuous power to maintain its contents.

NOR gate   A logic device where any one input or more having a 1-state will yield a 0-state output.

Off-phase clocking   Clocking signals that occur at some time other than when the normal active clock edge occurs. Typically, off-phase clocking is done at the midpoint between normal active clock edges.

One-shot (monostable multivibrator)   A digital circuit with two output states, one of which is stable and the other temporary. They can be triggered into the temporary state for a period of time determined by an associated RC network.

Operating speed   The speed at which a logic circuit or system must make decisions.

OR gate   A logic device where any one input or more having a 1-state is sufficient to produce a 1-state output.

Output levels   The actual output voltage level at the output terminals of a logic device.

Overshoot   When a signal goes beyond its normal range or steady-state level. When a digital signal goes below ground or above the power-supply level.

Parallel or split termination (sometimes described as a Thevenin's termination)   A termination network consisting of two series-connected resistors with a Thevenin's impedance (at the junction) with a prescribed relationship to the interconnecting line impedance. One of the resistors is connected to the supply voltage and the other to ground. The line is connected to the junction.

pc board   Printed circuit board.

Plane   A continuous layer or sheet of material in which any voids or cut-outs are small with respect to the wavelength of the highest frequency of interest.

Power-supply droop   The shift in device supply voltage caused by transient current flow through the inductance of the package power pins and other power-supply connections to the device.

Prototyping board   A universal circuit board without dedicated locations for circuit components and that can be used to quickly implement and test the operation of a new circuit.

Pull-down resistor A resistor connected to ground used to pull down unused inputs or lines that have the potential to float to an undefined logic level.

Pull-up resistor A resistor connected to the positive supply voltage used to pull up unused inputs or lines that have the potential to float to an undefined logic level.

Pull-down voltage A voltage used to provide a static *low* input level to unused inputs.

Pull-up voltage A voltage used to provide a static *high* input level to unused inputs.

Pulse width The time between the leading and trailing edge of a pulse.

Quiet time The time interval in a synchronous system between the time when all signals have settled until the next active clock edge.

Register A digital device (a flip-flop) used for temporary storage of digital data.

RESET or CLEAR input An input used to return a storage element to its logical 0-state.

RESET SIGNAL A signal used to initialize a system at power ON following a power transient or under other conditions.

Ringing When a signal overshoots and undershoots the final steady-state level a number of times following a logic level transition.

Rise time (t_r) The time interval required for a signal to transition from 10 to 90 percent of its final amplitude.

Schmitt trigger buffer A buffer with hysteresis between the positive-going and the negative-going input thresholds and that provides greater noise margin than conventional buffers.

SET input An input to a storage element that causes it to go to the logical 1-state.

Setup time (t_s) The time that the input data to a clocked logic device must be stable before the active (triggering) clock transition.

Short-circuit current (I_{SC}) The maximum output current which a device will provide into a short circuit or that is available to charge load capacitance.

Single-ended signals Signals with one discrete observable interconnection, i.e., a single wire where the current loop must be completed by the power or ground system.

Slew rate The rate of change of an output.

Solder clips A small clip used to make direct solder connections between stitch-wire pins and ground or power planes on stitch-wire boards with universal pin fields.

Solder washers A small washer used to make direct solder connections between wire-wrap pins and ground or **power planes on wire-wrap** boards with universal pin fields.

Source or series termination   An impedance in series with a line located near the driven end of the line that has a prescribed relationship to the interconnecting line characteristic impedance (Z_o).

Split termination   *See* Parallel termination.

State   The condition of a logic signal, i.e., whether it is a *high* or *low* level, or a logical 1 or logical 0.

Storage element   A flip-flop (see flip-flop).

Synchronous   All devices are clocked in step with a master clock signal.

Synchronous system   A digital system in which all sending and receiving clocked devices are clocked with a common clock frequency and phase.

Termination   Providing a source or load impedance that has a prescribed relationship to the interconnecting line impedance.

Thevenin's termination   *See* Parallel termination.

Three-state   Logic systems utilizing three states: defined *high* and *low* levels and an undefined high-impedance state.

Threshold voltage   The input voltage level at which the output logic level is no longer defined.

Transition time   The time required to transition between specified *high*-to-*low* or *low*-to-*high* logic levels.

Transmission line   An interconnecting signal line that is comparable to the wavelength of the signal frequency.

Transmission-line effects   Ringing, overshoot, undershoot, loading, and line propagation delays are effects of transmission-line interconnections. Transmission-line effects delay signal settling.

Transmission-line load   The characteristic impedance Z_o of a driven transmission line.

Undefined logic level or condition   The input or output level is not a defined *high* or *low* logic level.

Undershoot   When a digital signal rings back toward or across the nearest logic level threshold, i.e., when a digital signal transitions it may overshoot and then ring back or *undershoot*.

Unit load   Generally the load for one *high* or *low* input for a given TTL logic family, but in some cases the load for one input of the standard (original) TTL logic family (*high* and *low* unit loads are different).

Vias   Connections between layers in multilayer printed circuit boards.

Index